Questions & Answers: Condition and Use of Vehicles

B Adamson, M Jewell, J Lawton and J M Pugh

Croner Publications Ltd
Croner House
London Road
Kingston upon Thames
Surrey KT2 6SR
Telephone: 0181-547 3333

Copyright © 1993 Croner Publications Ltd
First published 1993
Second edition 1995

Published by
Croner Publications Ltd
Croner House
London Road
Kingston upon Thames
Surrey KT2 6SR
Tel: 0181-547 3333

While every care has been taken
in the writing and editing of this book,
readers should be aware that only Acts of Parliament
and Statutory Instruments have the force of law,
and that only the courts can authoritatively
interpret the law.

British Library Cataloguing in Publication Data
A CIP Catalogue Record for this book
is available from the British Library

ISBN: 1 85524 318 0

Printed by Clays Ltd, St Ives plc

THE AUTHORS

BOB ADAMSON is currently the Group Traffic Officer for TNT Express (UK) Ltd and was for many years the Accident Investigation Officer for the Lancashire Constabulary. During his years at the Lancashire Constabulary he was instrumental in organising courses on the law in relation to tachographs which were attended by representatives of police forces throughout the country and Department of Transport enforcement personnel.

MICHAEL JEWELL has been a writer and transport consultant for over 30 years, regularly contributing to journals such as *Commercial Motor, Motor Transport, Road Way, Freight* and *Headlight* on the subjects of road haulage licensing, the commercial and industrial use of all forms of transport and its legal aspects. For two years he was editor of the prestigious journal *Road Law* and is a regular contributor to various Croner publications.

JONATHAN LAWTON is a transport and industrial law specialist, a solicitor who has been practising since 1962. He has a haulage background; he was a director of a haulage company for several years. He is the Honorary Legal Advisor to the National Association of Warehousekeepers and a regular writer of articles regarding road transport law and industrial law matters. He is a regular contributor to various Croner publications.

JOHN MERVYN PUGH is a solicitor who was admitted in 1957. From 1961–84 he conducted all the prosecutions for the West Midlands Traffic Area in the County of Hereford and Worcester and the Metropolitan Borough of Sandwell. During this time he was for 13 years a part time clerk to the Alcester Justices and frequently sat as a Deputy Stipendiary Magistrate. He was appointed a Deputy Traffic Commissioner in 1984, took up the post as Traffic Commissioner for the Eastern Traffic Area in 1986 and from there was appointed Traffic Commissioner for the West Midlands and South Wales Traffic Areas; a position he still holds. He is a regular contributor to various Croner publications.

ACKNOWLEDGEMENT

The publishers wish to acknowledge their thanks to the Driver and Vehicle Licensing Agency at Swansea for their help and support.

INTRODUCTION

A vast amount of legislation affects the operation of "large vehicles" — goods vehicles and those carrying passengers. This revised edition of *Questions and Answers: Condition and Use of Vehicles* aims to provide answers to many of the most commonly asked questions that concern operators seeking to ensure that their operations are run legally. The answers have been compiled by a team of authors with a wealth of experience in the legalities of road transport operation. The areas covered include vehicle maintenance, licensing, taxation, insurance, health and safety, carriage of dangerous goods, environmental issues, tachographs and drivers' hours and speed limiters.

The book has been thoroughly revised to give the most up to date information and references, particularly in the areas of licensing and vehicle registration, health and safety and environmental issues. The authors and publisher are confident that the revised edition will prove as useful and popular as the original work.

Copies of legislation (Acts, Orders and Regulations) mentioned in the text may be ordered from any bookseller or obtained direct from HM Stationery Office. Publications concerning vehicle maintenance can be obtained from the Department of Transport Vehicle Inspectorate.

QUESTIONS & ANSWERS FOR ROAD TRANSPORT OPERATORS: CONDITION AND USE OF VEHICLES

MAINTENANCE, INSPECTIONS AND TESTING

Q1. Where can the law relating to operator licensing be found?

Q2. Which legislation covers maintenance?

Q3. Who has to be satisfied that there will be satisfactory facilities and arrangements?

Q4. Who appoints licensing authorities?

Q5. Do the Traffic Commissioners form part of the Department of Transport?

Q6. How many Traffic Commissioners are there?

Q7. Are Traffic Commissioners concerned with financial resources regarding maintenance?

Q8. How much money should be available for each vehicle?

Q9. What is an ECU?

Q10. What is the value of an ECU?

Q11. How can I find out the daily value of an ECU?

Q12. If the value of the ECU goes up or down, will this affect the amount of money that an operator has to have available?

Q13. What value will the Traffic Area Office accept when an application is made?

Q14. After a licence has been granted, can the Traffic Area Office summon the operator before the Traffic Commissioner to increase the amount of money available?

Q15. Is there any difference in the amount of money that has to be available for a Restricted Operator's Licence as opposed to a Standard National Licence or a Standard International Licence?

Q16. Is a bank overdraft acceptable?

Q17. What happens if the bank calls in the overdraft facility?

Q18. If the operator's financial position is in jeopardy, would the Operator's Licence be at risk?

Q19. What financial information has to be given to the Traffic Commissioner, if requested, when an application is made for a Standard Operator's Licence or the variation of such a licence?

Q20. After form GV79F has been completed, can the Traffic Commissioner require any further information?

1

Q21. What part does a financial assessor play in the granting of any licence?

Q22. If the financial assessor advises the Traffic Commissioner in writing, is the operator entitled to a copy of that advice?

Q23. Who decides how maintenance will be arranged?

Q24. What is a service?

Q25. How often should the vehicle be serviced?

Q26. Is a service the same as an inspection?

Q27. Should an inspection be time-based?

Q28. What does "time-based" mean?

Q29. Should planned maintenance take precedence over operations?

Q30. What is the *Inspection Manual*?

Q31. Why did the change take place from the Testers' Manual to the Inspection Manual?

Q32. Is the Inspection Manual available to anyone and, if so, where can copies be obtained from?

Q33. How much does the Inspection Manual cost?

Q34. What format is the Inspection Manual in?

Q35. On inspection sheets there is a column marked *TM number.* What is a TM number?

Q36. Does that mean that for TM number, we should read IM number?

Q37. Is it essential to have the Inspection Manual to inspect the vehicle?

Q38. Should every operator have an Inspection Manual?

Q39. Does this apply even if the maintenance is contracted out?

Q40. How should operators who contract out their maintenance use the Inspection Manual?

Q41. After the inspection, on presentation of the inspection sheet, what should operators check?

Q42. Is it important to have a certification on an inspection sheet to the effect that all the defects have been remedied and the vehicle is now in a fit and serviceable condition?

Q43. If an operator is called up before a Traffic Commissioner because of poor maintenance, would the fact that he or she had certification on an inspection sheet help the case?

Q44. Should operators have any other books besides the Inspection Manual?

Q45. How long have these books been available?

Q46. Are the books now available the same as those used by inspectors in the past?

Q47. Who should inspect vehicles?

Q48.   How can an operator join the Freight Transport Association?

Q49.   Is it advisable to have vehicles inspected by authorised dealers of the make of vehicle in question?

Q50.   Do all authorised dealers have mechanics available?

Q51.   How can an operator find out the address of the nearest authorised dealer?

Q52.   Is there any fixed, ideal planned maintenance?

Q53.   How should operators decide upon which system of planned maintenance to use?

Q54.   Can any plan be recommended?

Q55.   Is it possible to elaborate upon the system of planned maintenance?

Q56.   How good is the prescribed system of maintenance?

Q57.   If drivers have been employed for a long time, is it still necessary to provide them with letters setting out their duties?

Q58.   If only a few drivers are employed would it not be better to put a notice in the drivers' room or in the drivers' cabs?

Q59.   Why should drivers' reports be daily?

Q60.   Why should reporting books be personal to the driver? Would it not be better if they were personal to the vehicle?

Q61.   Why should sheets in the reporting book be consecutively numbered?

Q62.   How long should reporting books be kept for?

Q63.   Is it advisable to have a repair book?

Q64.   How can accountancy fees be saved by keeping a repair book?

Q65.   What is the time period of an inspection?

Q66.   Is it possible to say whether or not there is an average time for inspections to be carried out?

Q67.   If a vehicle is only used seasonally, how should the periodic times be arranged?

Q68.   Is it possible to conduct an MOT on a vehicle even though there is life in the current MOT certificate?

Q69.   Which certificate is then the live one?

Q70.   Is it possible to obtain a convenient appointment at a testing station?

Q71.   What form does the test take?

Q72.   With vehicles travelling over rough terrain, is it advisable to have them inspected more frequently?

Q73.   With busy operators how is it possible to have vehicles inspected without losing operational time?

Q74.   What is a *transport manager*?

Q75.   Is a transport manager restricted to a certain number of vehicles?

Q76.   Does every operator have to have a transport manager?

Q77.   Does a Standard National or Standard International Licence operator have to have a transport manager?

Q78.   How does an operator decide that a part time transport manager is what is required?

Q79.   Does a transport manager have to have specific qualifications?

Q80.   What is a Certificate of Professional Competence?

Q81.   How can a prospective transport manager find out about the examinations?

Q82.   Can a transport manager lose his or her "good repute"?

Q83.   Can a transport manager appeal against the decision to remove his or her "good repute"?

Q84.   What is the Transport Tribunal and where is it based?

DRIVER LICENSING

Q85.   What is the main legislation governing driver licensing for large goods vehicle (LGV) and passenger carrying vehicle (PCV) drivers?

Q86.   What was the major change so far as the licence document itself is concerned?

Q87.   How do the previous groups equate with the new categories of entitlement?

Q88.   Do the new categories show any specific change and, if so, what?

Q89.   Now that there is one licence, does this mean that every LGV/PCV driver has to have a new style one?

Q90.   Were there any changes in the driving test following 1 April 1991?

Q91.   What are the addresses and telephone numbers of the DSA to whom an application should be made for an LGV or PCV driving test?

Q92.   Is an LGV entitlement the same as a previous HGV licence?

Q93.   Is a PCV entitlement the same as an old style PSV licence?

Q94.   When heavy goods vehicle (HGV) licences became described as large goods vehicle (LGV) entitlements, and public service vehicle (PSV) licences became passenger carrying vehicle (PCV) entitlements, why is it that LGVs are still referred to as HGVs and PCVs as PSVs?

Q95.   Will there be a later change in this nomenclature?

Q96.   At what age can the holder of an ordinary driving licence apply for an LGV or a PCV entitlement?

Q97.   When driving under a provisional LGV licence, does the driver have to carry "L" plates?

Q98.   What are the entitlements of minibus drivers following 1 April 1991?

Q99.   Can the holder of a full ordinary driving licence obtained prior to 1 April 1991 drive a bus which is not in service?

Q100.   Do the new regulations affect drivers who were described as "movers" prior to 1 April 1991 and fitters who drive vehicles without passengers in order to test the vehicle?

Q101.   What are "Grandfather Rights"?

Q102.   Can "Grandfather Rights" be acquired for any new entitlement?

Q103.   What is the position with regard to a licence required to drive a goods vehicle which is within the Operator's Licence and is over 3500kg but under 7500kg?

Q104.   Prior to 1 April 1991 the issuing and control of vocational licences was dealt with by the Traffic Area offices. What is their function now in relation to the issuing and conduct of vocational driver entitlement?

Q105.   As the Traffic Commissioners are agents for the Secretary of State for Transport, who is the licensing authority now?

Q106.   Is the Secretary of State for Transport bound by any decision that the Traffic Commissioner may make?

Q107.   What is the position with regard to medical examinations and further periodic medical examinations required in order to maintain an LGV or PCV entitlement?

Q108.   What fees are payable in respect of the medical examination?

Q109.   Can an applicant be refused a vocational entitlement on medical grounds?

Q110.   What medical grounds should drivers be aware of which could bar them from holding LGV or PCV licences?

Q111.   Does a driver have the right to appeal against the medical decision?

Q112.   If the appeal is unsuccessful in the magistrates' court, can the DVLA apply for costs against the driver who has appealed?

Q113.   Does the driver have to pay for his or her own medical evidence, which he or she brings to refute the evidence given by the DVLA doctors?

Q114.   Can the driver apply for Legal Aid in respect of such an application?

Q115.   Are there any other methods of obtaining financial assistance?

Q116.   What is the medical report form D4 (previously DTp 20003) and is the driver entitled to see it?

Q117.   What special requirements are there for a medical examination and submission of a medical form D4?

Q118.   Where should a driver direct any enquiries relating to medical matters?

Q119.   Is it always necessary to give a date of birth and postcode when communicating with the DVLA?

Q120. Is it true that each driving licence number is based on the holder's date of birth?

Q121. Which department at the DVLA should drivers communicate with about their driving licences?

Q122. If a licence is lost, can a duplicate be obtained with the same entitlements upon it?

Q123. What is the cost of a duplicate licence?

Q124. What are the costs of an ordinary licence and provisional LGV and PCV licences?

Q125. What are the costs of the various driving tests?

Q126. Are test fees payable in advance?

Q127. When the test is taken will the examiner require to see the provisional licence?

Q128. Are examiners insured for taking tests, or do they require to see the insurance of the vehicle in which a driver is taking a test?

Q129. If a motoring offence is committed and penalty points are endorsed on the counterpart to the harmonised licence, does this affect the lorry/bus entitlement?

Q130. If there are penalty points endorsed on a car licence and an application is made for a provisional PCV or LGV, is this looked at by the Traffic Commissioners?

Q131. How long does a provisional licence last and can a new application be made for renewal after its expiry?

Q132. Is there any period in which a test must be taken after the grant of a provisional licence?

Q133. If an LGV driver is convicted of motoring offences which are not endorsable, does that affect his or her holding of the lorry/bus entitlement?

Q134. If a PCV driver commits offences, other than motoring offences, can they in any way affect his or her holding of a PCV licence?

Q135. What is the difference between offences committed by LGV and PCV drivers?

Q136. When offences are committed which can affect the vocational entitlement, how is the driver notified of further action?

Q137. While the magistrates' courts or the Crown Court deal with motoring offences why do they not also deal with offences relating to vocational entitlement?

Q138. If the offender is a HGV or PSV driver and is still in possession of the old licence, does that licence have to be handed to the magistrates at the time of endorsement or disqualification?

Q139. If the entitlement is on the new harmonised licence, rather than the old style licence, does this still mean that the magistrates can only take action against the car entitlement?

Q140. If a driver is disqualified from holding a car driving licence as a result of committing an offence, does this automatically revoke any lorry or bus entitlement?

Q141. Can an applicant apply for a provisional car licence and at the same time for a provisional LGV/PCV licence?

Q142. Can action be taken by a Traffic Commissioner when any penalty points are endorsed upon the car licence?

Q143. Is it true that any endorsable offence also carries with it a possible disqualification?

Q144. Under the "totting up" procedure in magistrates' courts, when a driver has 12 or more penalty points, is the driver automatically disqualified?

Q145. Does the Secretary of State have a separate points system or a number of points when action is taken against the LGV/PCV entitlement?

Q146. What form does such a warning take?

Q147. Does this mean that if an offence is committed in a private motor car it can affect the LGV/PCV entitlement?

Q148. Can the LGV/PCV entitlement be revoked whilst the car entitlement remains and what vehicles, if any, can then be driven?

Q149. Is it true that an LGV entitlement may, in the future, be required for driving a vehicle in excess of 3500kg?

Q150. If the LGV/PCV entitlement is revoked, when can a new application be made?

Q151. Is revocation generally accompanied by a period of disqualification?

Q152. Is there any set period of disqualification in relation to any offences?

Q153. If a Traffic Commissioner wishes to take action against a driver's LGV/PCV entitlement does he or she first of all notify the holder of this fact?

Q154. Do Traffic Commissioners have power to call the holder of a LGV/PCV entitlement before them?

Q155. After a Traffic Commissioner has made a decision, what right of appeal is there?

Q156. Is any decision the Traffic Commissioner makes stayed, pending the hearing of an appeal?

Q157. Is any stay application bound to succeed?

Q158. Do appeals generally succeed?

Q159. What are the chances of success in a medical case appeal before magistrates?

Q160. Can an appellant insist upon a medically qualified person sitting on the bench of magistrates presiding over the case?

Q161. Does a magistrates' court have the power to award costs?

Q162. Can a driver with a LGV/PCV entitlement request a meeting with the Traffic Commissioner to discuss his or her penalty points?

Q163. What form do hearings before the Traffic Commissioners take?

Q164. If the driver is seen privately by the Traffic Commissioner is there any variation in the procedures?

Q165. If a suspension is ordered can it be said to commence at a future date or must it be immediate?

Q166. What sort of record is taken at public inquiries?

Q167. Is the oath given at these public inquiries?

Q168. What applications are heard by a Traffic Commissioner?

Q169. What is the position of a driver who has previously been disqualified for a drink-related offence?

Q170. When a driver has been disqualified and has LGV/PCV entitlement, how can he or she restore that LGV/PCV entitlement when his or her car licence is returned?

Q171. Are there any set rules as to the restoration of the LGV/PCV entitlement?

Q172. What happens if a driver has been disqualified more than once for drink-driving related convictions?

Q173. To whom is the Traffic Commissioner responsible in making such decisions?

Q174. Can members of the public report bad driving incidents?

Q175. What action will be taken?

Q176. If just the registered number of the vehicle is given, is that sufficient to trace it?

Q177. Can complaints against drivers be made anonymously?

Q178. What powers does the Traffic Commissioner in fact have over an LGV/PCV driver's entitlement?

Q179. Are warnings taken into consideration if further offences are committed?

Q180. How long does the LGV/PCV entitlement last?

Q181. If a licence is revoked or suspended, is any part of the fee returned?

Q182. How long do endorsements stay on an ordinary driving licence?

Q183. If a driver sends an application to the DVLA to renew an existing licence, is he or she covered to drive?

Q184. What should a driver do on changing his or her name or address?

Q185. When a licence holder dies, what action should the next of kin take?

Q186. Are any special licences required for the driving of historic goods vehicles?

Q187. How is a historic goods vehicle defined for driving licence purposes?

TAXATION

Q188. What is the legislation governing the taxation of vehicles?

Q189. How do I apply for a tax disc?

Q190. Does a tax disc in the window mean that the vehicle has been taxed?

Q191. When the tax expires do I have any leeway period in which to obtain the new tax?

Q192. My vehicle is taxed at the concessionary rate for 38,000kg on six axles. Can I pull a two-axle trailer with it?

Q193. My vehicle is taxed at 38,000kg on six axles. Can I "lift" an axle?

Q194. I have recently taxed a vehicle, but it will now be off the road for some time. Can I use the disc on another vehicle?

Q195. My tax has run out and I will not be able to tax the vehicle for a few days. What should I do with the existing disc?

Q196. Does the receipt of the money for the tax confirm the type of vehicle that has been taxed?

Q197. Are any vehicles exempt from tax?

Q198. I have a high pressure water-jetting vehicle and I understand that it is not a goods vehicle. Is this correct?

Q199. What is "revenue weight"?

Q200. How would the duty be calculated for my high pressure water-jetting vehicle?

Q201. I have been told that my vehicle is not a goods vehicle, but the authorities refuse to tax it unless I tender it as a goods vehicle. What should I do?

Q202. I want to down rate my vehicle. Can I reclaim the difference in duty?

Q203. My vehicle has been off the road untaxed for some time. Will I have to pay back duty to the point at which it was last taxed?

Q204. My vehicle has a mechanical fault and is temporarily parked at the side of my house. Do I have to tax it?

Q205. I have paid for the new tax but the disc has not arrived. Can I alter my old one?

Q206. I have an articulated tractor plated to 38,000kg which will pull engineering plant. Do I need to tax it at 38,000kg?

Q207. I used a special types vehicle to bring a small piece of equipment back for my own use. Is this unlawful?

Q208. I have a spare vehicle. If I get a trade licence can I use this vehicle if one of my taxed vehicles is broken-down?

Q209. I have been offered a disc that still has time to run at a price that will save me money. Should I buy it?

Q210. My disc was accidentally destroyed. I have applied for a new disc — can I run the vehicle in the meantime?

Q211. I have applied for a tax disc and paid for it, but it has not arrived. Can I use the vehicle?

Q212. We are thinking of offering a combined road/rail service. Are there any tax benefits?

INSURANCE

Q213. Where can the law relating to motor insurance be found?

Q214. In applying for insurance cover what has to be disclosed?

Q215. If drivers have convictions on their ordinary driving licence should these be disclosed?

Q216. If a particular class of insurance is refused by an insurance company can I apply to another company?

Q217. Is a driver's driving entitlement relevant to insurance?

Q218. Should I read all the small print on my policy?

Q219. What is a "traders' policy"?

Q220. If a premium is not paid what happens to the policy?

Q221. Can a limited company be penalised in the same way?

Q222. Is it advisable to seek the help of an insurance broker in arranging insurance?

Q223. As all motor vehicles have to be insured by law what is the minimum requirement that the law demands?

Q224. What are "third party risks"?

Q225. How do I obtain insurance cover for repairs following an accident to my own vehicle?

Q226. How do I decide between third party and comprehensive cover?

Q227. How could I reduce my premiums?

Q228. How can I recover the uninsured losses?

Q229. What do I do if the other party ignores my letters?

Q230. How do I issue a summons at the County Court?

Q231. What is the procedure in a small claims court?

Q232. What is "goods in transit insurance"?

Q233. How are customers made aware of a carrier's conditions?

Q234. Is wider cover available?

Q235. Can an insurance company impose additional conditions?

Q236. What advice can be given to a driver in respect of vehicle and load security?

Q237. Do I need passenger liability insurance?

Q238. What must a driver do after being involved in an accident?

Q239. What is a "specified animal"?

Q240. What is "roadside property"?

Q241. How do I go about making a claim?

HEALTH AND SAFETY

Q242. What is the "Six Pack"?

Q243. Where can I obtain a copy of the "Six Pack"?

Q244. Is there an overall guide to the regulations?

Q245. How do these regulations affect me as an employer?

Q246. I am self-employed. Am I affected by the regulations?

Q247. What are the "COSHH" regulations? Are they the same as the "Six Pack"?

Q248. I have a small company. Do I need to worry about all these regulations?

Q249. I understand that assessments of risk have to be carried out. Who would be best suited to undertake these?

Q250. Will my insurance company help regarding assessments of risk and, if I am insured, do I need to bother with assessments in any case?

Q251. One of my staff is pregnant. Is she or her child in any danger if she continues to use a VDU?

Q252. I am about to alter the building in which we work. Are there any problems I should be aware of?

Q253. How is "workplace" defined?

Q254. Will I need professional help to ensure that I am complying with the workplace regulations?

Q255. A number of the people I employ have been with me a long time and are totally familiar with the tools they are using. Do I need to train them?

Q256. What are "tools" for the purposes of the regulations?

Q257. The machine that I have just bought is provided with guards. Do I have any additional duty?

Q258. We make the workers responsible for ensuring that the tools they use are in good condition. Should we confirm that instruction in writing?

Q259. I have done my manual handling assessments, but cannot control my drivers while they are away from the premises. Would I be responsible if a driver was injured?

Q260. I have a female driver. Should I ask her if she is pregnant as she will have to lift some weight?

Q261. What are considered to be the driver's responsibilities with regard to ensuring the safety of the vehicle?

Q262. If an employee who has been with me for less than two years leaves, or is dismissed, claiming it was for health and safety reasons, can he or she take me to an industrial tribunal?

CARRIAGE OF DANGEROUS GOODS

Q263. Where can the law, in relation to the carriage of dangerous goods, be found?

Q264. What are the main responsibilities of the operator when dangerous goods are being carried?

Q265. What are the main responsibilities of drivers when dangerous goods are being carried?

Q266. What substances are covered by the regulations?

Q267. Is there a laid down definition of the term "dangerous substance"?

Q268. What is the "Approved List"?

Q269. Where can I obtain copies of the Approved Lists?

Q270. Is it an offence not to comply with the Approved Code of Practice (ACOP)?

Q271. What is the definition of "explosives"?

Q272. What is meant by "conveyance by road"?

Q273. What is meant by "carriage" with regard to explosives?

Q274. Do substances always require labelling?

Q275. What are the main features of the supply label in general?

Q276. What are the main features of the label for conveyance by road?

Q277. Who is responsible for complying with the law when dangerous goods or explosives are being carried?

Q278. Who is regarded as the operator?

Q279. Are the terms "road tanker" and "tank container" defined?

Q280. Do road tankers, tank containers and vehicles used for the carriage of dangerous goods and explosives have to comply with any specific requirements?

Q281. What are the maximum quantities of explosives that can be carried?

Q282. What are the testing and certification requirements for road tankers, tank containers and vehicles used to carry dangerous goods and explosives?

Q283. When do the Road Traffic (Carriage of Dangerous Substances in Road Tankers and Tank Containers) Regulations 1992 (SI 1992 No.743) not apply?

Q284. What information should be available before dangerous goods or explosives are carried?

Q285. How should such information be treated once unloading is completed?

Q286. When do the Packaged Goods Regulations (SI 1992 No.742) apply?

Q287. When do the Road Traffic (Carriage of Dangerous Substances in Packages etc) Regulations 1992 (SI 1992 No.742) not apply?

Q288. What precautions should be taken during loading, stowage and unloading?

Q289. What precautions should be taken against fire or explosion?

Q290. When must vehicles carrying dangerous goods in packages be marked?

Q291. What marking is required for vehicles carrying explosives?

Q292. What are the rules on overnight parking?

Q293. What are the rules for short stops?

Q294. What steps should be taken when vehicles are carrying explosives?

Q295. Are there any restrictions on the routes taken by vehicles carrying explosives, on the duration of the journey and delivery?

Q296. Are there any minimum age requirements for drivers when vehicles are carrying explosives?

Q297. What are the training requirements for drivers and attendants of vehicles carrying dangerous goods and explosives?

Q298. What are the current training requirements under the Road Traffic (Carriage of Dangerous Substances in Packages etc) Regulations 1992 (SI 1992 No.742)?

Q299. How can drivers obtain the required ADR training certificates?

Q300. What are the current minimum ADR Directive requirements for tanker driver training?

Q301. What are the driver's responsibilities in relation to ADR certificates?

Q302. What action should drivers take in the event of an incident/accident involving dangerous goods?

ENVIRONMENTAL ISSUES

Q303. Where can the environmental provisions relating to goods vehicle operators' licensing be found?

Q304. What is the aim of the environmental provisions in the operators' licensing legislation?

Q305. What is an "operating centre"?

Q306. What does a licensing authority examine when considering a licence application under the environmental provisions?

Q307. In granting an application, what environmental conditions can a licensing authority impose on a licence?

Q308. Is there any right of appeal against a decision of a licensing authority to impose conditions or refuse an application on environmental grounds?

Q309. What happens if the conditions imposed are breached?

Q310. Do conditions imposed apply to every vehicle using the operating centre?

Q311. Who can object to a licence application?

Q312. Can local residents object to a licence application?

Q313. What procedure should objectors and representers follow?

Q314. Does the newspaper have to be a local one?

Q315. What information has to be given regarding the environmental aspect when making a licence application?

Q316. What does the licensing authority regard as the "vicinity" of an operating centre?

Q317. If objections and/or representations are made and are held to be valid will there be a public inquiry?

Q318. What is "controlled waste"?

Q319. Do I have to be registered to carry controlled waste?

Q320. Are there any exceptions to the registration?

Q321. What is the duty of care required when carrying controlled waste?

Q322. Is there any other documentation required for the carrying of waste, apart from the written description of waste?

Q323. For how long should records relating to controlled waste be kept?

Q324. Where can controlled waste be deposited?

Q325. What is "special waste"?

Q326. What is the documentation required for the carrying of special waste?

Q327. For how long should records relating to special waste be kept?

Q328. What is "transfrontier shipment"?

Q329. What are the rules and the documentation required for the carrying of hazardous waste?

Q330. Where can I find the legislation relevant to water pollution?

Q331. What is the position regarding possible water pollution with regard to vehicle maintenance and washing off operations?

Q332. What steps can be taken to avoid water pollution?

Q333. What are the penalties for polluting a watercourse?

TACHOGRAPHS AND DRIVERS' HOURS

Q334. Where can the law relating to EC drivers' hours and tachograph legislation be found?

Q335. Where can the rules regarding domestic drivers' hours be found?

Q336. What is the purpose of the legislation?

Q337. When do the domestic drivers' hours rules apply?

Q338. Are there any exemptions to the domestic drivers' hours rules?

Q339. Is there any requirement for drivers to keep records under the domestic drivers' hours rules?

Q340. Are there any exemptions from keeping records under the domestic drivers' hours rules?

Q341. What happens if the domestic drivers' hours rules are contravened?

Q342. What is the position when drivers drive partly under the EC drivers' hours rules and partly under the domestic drivers' hours rules?

Q343. What vehicles are subject to the tachograph regulations?

Q344. Are there any exemptions from the tachograph legislation?

Q345. What are the employer's responsibilities regarding drivers' hours and tachograph regulations?

Q346. What happens in circumstances where a required vehicle does not have a tachograph fitted or where a tachograph is not being used in accordance with the regulations?

Q347. Is there any duty on employers to check the tachograph charts?

Q348. What are the duties placed on drivers?

Q349. When drivers recommence driving after a holiday, must they carry with them the chart covering the last day on which they drove before the holiday?

Q350. What happens when drivers are away from their vehicle and unable to operate the tachograph?

Q351. Does other work after a driver has finished driving have to be recorded on the tachograph chart?

Q352. What information do drivers have to enter on the centre field of a tachograph chart?

Q353. How often should tachographs be inspected and calibrated?

Q354. What happens when a tachograph is faulty or the seals are broken?

Q355. What are the daily driving limits under EC and domestic rules?

Q356. How long can drivers drive under the EC rules before they have to take a break?

Q357. Is there a limit on the number of hours drivers can drive in a fortnight under the EC rules?

Q358. What is the definition of a working "day" and a working "week"?

Q359. What are the daily rest requirements under the EC rules?

Q360. What are the weekly rest requirements under the EC rules?

Q361. What happens when drivers are on a ferry boat or train?

Q362. What happens in emergencies?

Q363. Do the EC regulations prohibit bonus payments to drivers?

SPEED LIMITERS

Q364. Where can the law in relation to speed limiters be found?

Q365. What are the main provisions of EC Directive 6/1992?

Q366. Are there any exceptions to the EC rules?

Q367. Which vehicles (under UK law) require speed limiters to be fitted?

Q368. Which vehicles are exempt?

Q369. Do speed limiters have to meet a particular specification?

Q370. What are the requirements of British Standard BS AU 217?

Q371. How can you tell whether or not a vehicle has been fitted with a speed limiter?

Q372. Who is regarded as an "authorised sealer"?

Q373. Is it an offence to drive a vehicle on the road when the speed limiter is not functioning?

MAINTENANCE, INSPECTIONS AND TESTING

Q1. **Where can the law relating to operator licensing be found?**

A. In the **Transport Act 1968** (as amended).

The **Goods Vehicles (Licensing of Operators) Bill** is expected to be enacted before the Parliamentary Summer Recess of 1995. When enacted, it will consolidate the **Deregulation and Contracting Out Act 1994** with Part V of the **Transport Act 1968**, as amended.

New regulations will also be laid before Parliament and it is anticipated that they, and the **Goods Vehicles (Licensing of Operators) Act 1995**, will come into force on 1 January 1996. However, there will be no change in maintenance requirements.

The new regulations will replace the **Goods Vehicles (Operators' Licences, Qualifications and Fees) Regulations 1984** (SI 1984 No.176) and all subsequent amendments, which will be repealed, as will Part V of the **Transport Act 1968** (as amended). Until Commencement Orders are made, the **Transport Act 1968** (as amended) remains the authority.

Q2. **Which legislation covers maintenance?**

A. Section 64(2)(d) of the **Transport Act 1968** states that there should be satisfactory facilities and arrangements for maintaining the authorised vehicles in a fit and serviceable condition.

Q3. **Who has to be satisfied that there will be satisfactory facilities and arrangements?**

A. The licensing authority, whose general title is the Traffic Commissioner. He or she is the *licensing authority* for goods vehicle operators and the *Traffic Commissioner* for PSV operators and driver licensing. As Traffic Commissioner he or she acts as agent for the Secretary of State.

Q4. **Who appoints licensing authorities?**

A. They are appointed under statute by the Secretary of State for Transport.

Q5. **Do the Traffic Commissioners form part of the Department of Transport?**

A. No. They are independent of the Department of Transport.

Q6.   How many Traffic Commissioners are there?

A.   At present there are seven, but there are eight separate Traffic Areas (the Traffic Commissioner for West Midlands also serves South Wales). The addresses and telephone numbers of the eight Traffic Areas are listed below.

Scottish Traffic Area Office, 83 Princes Street, Edinburgh EH2 2ER. Tel: 0131-225 5494.

North Eastern Traffic Area Office, Hillcrest House, 386 Harehills Lane, Leeds LS9 6NF. Tel: 0113 283 3533.

North Western Traffic Area Office, Portcullis House, Seymour Grove, Manchester M16 0NE. Tel: 0161-886 4000.

West Midlands Traffic Area Office, Cumberland House, 200 Broad Street, Birmingham B15 1TD. Tel: 0121-608 1060.

Eastern Traffic Area Office, Terrington House, 13–15 Hills Road, Cambridge CB2 1NP. Tel: 01223 358922.

South Wales Traffic Area Office, Caradog House, 1–6 St. Andrew's Place, Cardiff CF1 3PW. Tel: 01222 394027.

Western Traffic Area Office, The Gaunt's House, Denmark Street, Bristol BS1 5DR. Tel: 0117 975 5000.

South Eastern and Metropolitan Traffic Area Office, Ivy House, 3 Ivy Terrace, Eastbourne BN21 4QT. Tel: 01323 721471.

Q7.   Are Traffic Commissioners concerned with financial resources regarding maintenance?

A.   Yes, under s.64(2)(e) of the 1968 Act, the licensing authority has to be satisfied that the provision of maintenance facilities and arrangements will not be prejudiced by reason of the applicant's insufficient financial resources for that purpose.

Q8.   How much money should be available for each vehicle?

A.   This is not specified for restricted and national vehicles. Schedule 6 of the **Goods Vehicles (Operators' Licences, Qualifications and Fees) Regulations 1984** (SI 1984 No.176) states that the applicant should have "available sufficient financial resources to ensure the establishment and proper administration of the road transport undertaking".

For international operators, capital and reserves of 3000 ECUs per vehicle, or 150 ECUs per tonne (whichever is the less), are required.

Q9. **What is an ECU?**

A. An ECU is a European Currency Unit.

Q10. **What is the value of an ECU?**

A. ECUs fluctuate according to the European Monetary System.

Q11. **How can I find out the daily value of an ECU?**

A. By studying the financial pages of the national/ international press, eg the *Financial Times.*

Q12. **If the value of the ECU goes up or down, will this affect the amount of money that an operator has to have available?**

A. Yes, because the ECU's value fluctuates, as do all currency values.

Q13. **What value will the Traffic Area Office accept when an application is made?**

A. The current value of the ECU on the day the application is accepted.

Q14. **After a licence has been granted, can the Traffic Area Office summon the operator before the Traffic Commissioner to increase the amount of money available?**

A. It is doubtful. No case has yet been decided, but it is fair to say that the date that the application is accepted is the relevant date.

Q15. **Is there any difference in the amount of money that has to be available for a Restricted Operator's Licence as opposed to a Standard National Licence or a Standard International Licence?**

A. Yes. Generally, for a Restricted Operator's Licence the Traffic Commissioner will wish to know whether the applicant has sufficient funds to maintain the vehicle. For a Standard National or Standard International Licence the Traffic Commissioner will need to be satisfied that there are sufficient funds to run the business; this of course includes the maintenance of the vehicles.

Q16. **Is a bank overdraft acceptable?**

A. Yes, provided that a letter is produced from the bank indicating that overdraft facilities are available.

Q17. **What happens if the bank calls in the overdraft facility?**

A. The financial position of the operator would then be in jeopardy.

Q18. **If the operator's financial position is in jeopardy, would the Operator's Licence be at risk?**

A. Yes. Under s.69(1)(e) of the **Transport Act 1968** (as amended) the licensing authority which granted an Operator's Licence may direct that it be:

- revoked
- suspended
- terminated on a date earlier than that on which it would otherwise expire under s.67 of the **Transport Act 1968** (as amended), or
- curtailed

because there has been, since the licence was granted or varied, a *material change* in any of the circumstances of the holder of the licence which were relevant to the grant or variation of the holder's licence. Therefore it follows that a change in financial circumstances becomes a material change, as that was one of the matters that was taken into consideration and was relevant to the grant or variation of the licence.

Q19. **What financial information has to be given to the Traffic Commissioner, if requested, when an application is made for a Standard Operator's Licence or the variation of such a licence?**

A. The financial information requested on form GV79F. This form has to be completed and submitted with each application.

Q20. **After form GV79F has been completed, can the Traffic Commissioner require any further information?**

A. Yes. If the Traffic Commissioner is not satisfied, or there is some doubt as to the accuracy of any information given, he or she can call upon one of the financial assessors to the Secretary of State for Transport to give advice as to whether or not there are sufficient funds.

Q21. **What part does a financial assessor play in the granting of any licence?**

A. The financial assessor can be called upon by the Traffic Commissioner to advise verbally or in writing. However, if the Traffic Commissioner is still not satisfied and decides to call a public inquiry, the financial assessor can sit with the Traffic Commissioner at the public inquiry and can ask questions of the applicant.

Q22. **If the financial assessor advises the Traffic Commissioner in writing, is the operator entitled to a copy of that advice?**

A. The Transport Tribunal have ruled that where financial standing remains an issue, or there is an objection on financial grounds, a licensing authority must disclose any report prepared by a financial assessor to the licence applicant and any objector must be given an opportunity of commenting on the report before a decision is reached. However, if a licensing authority feels it necessary to grant a licence, or to refuse it on grounds other than financial grounds, and there is no objection on financial grounds, there is no need for the licensing authority to disclose the report or to ask for comments upon it.

Q23. **Who decides how maintenance will be arranged?**

A. The operator, but the Traffic Commissioner must be satisfied that the facilities and arrangements for maintaining the authorised vehicles in a fit and serviceable condition are satisfactory.

Q24. **What is a service?**

A. A service is when the vehicle is submitted to a mechanic for the changing of oils and filters and the checking of all matters prescribed in the manufacturer's handbook.

Q25. **How often should the vehicle be serviced?**

A. The vehicle should be serviced in accordance with the manufacturer's handbook. Manufacturers normally recommend that vehicles should be serviced when they have covered a certain number of miles or at periodic time intervals, eg every 6000 miles or six months, whichever happens first.

Q26. **Is a service the same as an inspection?**

A. No. The two are entirely different. An *inspection* is a safety inspection when the vehicle is placed over a pit or on a ramp and all items listed in the *Inspection Manual* are checked.

Q27. **Should an inspection be time-based?**

A. Yes. The *Guide to Roadworthiness*, published by the Department of Transport in association with the Freight Transport Association, Road Haulage Association and the Bus and Coach Council (now the Confederation of Passenger Transport), clearly recommends that all inspections should be time-based. The guide is available from HMSO and currently costs £3. It is expected that a new, revised edition of the guide will be issued late in 1995 or early 1996.

Q28. **What does "time-based" mean?**

A. *Time-based* means that a fixed date should be planned for the inspection of the vehicle and that the inspection should take place on that date.

Q29. **Should planned maintenance take precedence over operations?**

A. Yes. All operations must be planned around the maintenance.

Q30. **What is the *Inspection Manual*?**

A. This used to be referred to as the *Testers' Manual*, but is now called the Inspection Manual.

Q31. **Why did the change take place from the Testers' Manual to the Inspection Manual?**

A. Originally, before the Vehicle Inspectorate became a separate agency, a publication entitled the *Heavy Goods Vehicle Testers' Manual* was used for the testing of vehicles. Now that the Vehicle Inspectorate is a government agency and its publications have been made available to all operators, the name has been changed to the *Heavy Goods Vehicle Inspection Manual*, commonly called the Inspection Manual.

Q32. **Is the Inspection Manual available to anyone and, if so, where can copies be obtained from?**

A. Yes. Anyone can get a copy from HMSO outlets, testing stations and any office of the Vehicle Inspectorate.

Q33. **How much does the Inspection Manual cost?**

A. £15.

Q34. **What format is the Inspection Manual in?**

A. It is in loose-leaf format; updates can be obtained on a standing order basis. It is anticipated that the cost of each replacement page will be £2.

Q35. **On inspection sheets there is a column marked *TM number*. What is a TM number?**

A. A TM number refers to the page numbers of the original Testers' Manual. These page numbers are exactly the same as in the Inspection Manual.

Q36. **Does that mean that for TM number, we should read IM number?**

A. Yes. Eventually all inspection sheets, when reprinted, will refer to an IM number.

Q37. **Is it essential to have the Inspection Manual to inspect the vehicle?**

A. Yes. The Inspection Manual sets out numerically all the items that have to be inspected, giving the method of inspection and the reason for rejection.

Q38. **Should every operator have an Inspection Manual?**

A. Yes.

Q39. **Does this apply even if the maintenance is contracted out?**

A. Yes, because operators can contract out their maintenance, but they cannot contract out their responsibilities.

Q40. **How should operators who contract out their maintenance use the Inspection Manual?**

A. Any operators who have contracted out their maintenance should make quite certain that the outside contractors know how to inspect the vehicle. It is recommended that operators should watch an inspection taking place and should have with them a copy of the Inspection Manual.

Q41. **After the inspection, on presentation of the inspection sheet, what should operators check?**

A. They should check the inspection sheet against the Inspection Manual, noting the items that failed and the fact that they have been remedied.

Q42. **Is it important to have a certification on an inspection sheet to the effect that all the defects have been remedied and the vehicle is now in a fit and serviceable condition?**

A. Yes. This is protection for operators, because a mechanic who has inspected the vehicle has put a signature to the work.

Q43. **If an operator is called up before a Traffic Commissioner because of poor maintenance, would the fact that he or she had certification on an inspection sheet help the case?**

A. In all probability, yes, because whilst the maintenance responsibility cannot be contracted out, the operator could be deemed, with such certification, to have done all that was reasonably possible to ensure the vehicle was in a fit and serviceable condition. Furthermore, the operator would be in a contractual relationship with the outside contractors, and if the latter had failed to comply with the contract and/or had falsely completed an inspection sheet, the operator would have a legal remedy in the civil courts for damages.

Q44.   Should operators have any other books besides the Inspection Manual?

A.   Yes — *The Categorisation of Defects* (£18), *The Amplification of Defects* (£18) and *The Guide to Roadworthiness* (£3) (see Question 27), which are available from HMSO outlets, testing stations and Vehicle Inspectorate offices.

Q45.   How long have these books been available?

A.   Since the Vehicle Inspectorate became a government agency in April 1991. Before that, the books were not available to operators because they were the personal books of vehicle inspectors.

Q46.   Are the books now available the same as those used by inspectors in the past?

A.   No. They include all the information that was available previously to vehicle inspectors, but they have been expanded to cover other matters and are now thorough and complete.

Q47.   Who should inspect vehicles?

A.   Inspections can either be contracted out or operators can employ their own fitters. Alternatively, the Freight Transport Association has a maintenance service, which is available to members.

Q48.   How can an operator join the Freight Transport Association?

A.   By contacting the Freight Transport Association either at its Head Office at Hermes House, St. John's Road, Tunbridge Wells, Kent TN4 9UZ (tel: 01892 526171) (this is also the regional office for the London and South Eastern region) or at one of the regional offices listed below.

Midlands Region: Hermes House, 20 Coventry Road, Cubbington, Leamington Spa CV32 7JN. Tel: 01926 450020.

Northern Region: Springwood House, Low Lane, Horsforth, Leeds LS18 5NU. Tel: 0113 258 9861.

Scottish Region (including Northern Ireland): Hermes House, Melville Terrace, Stirling FK8 2ND. Tel: 01786 471910 (Northern Ireland: 01232 241616).

Western Region: Hermes House, Queen's Avenue, Clifton, Bristol BS8 1SE. Tel: 0117 973 1187.

Q49. **Is it advisable to have vehicles inspected by authorised dealers of the make of vehicle in question?**

A.   Yes, because since the mid 1980s engines have become increasingly sophisticated electronically. Each authorised dealer of each manufacturer has to submit its staff for detailed training concerning the engine, mechanical parts and the vehicle as a whole. It is strongly recommended, therefore, that if your vehicle is a Mercedes, for example, it should be inspected by a Mercedes dealer.

Q50. **Do all authorised dealers have mechanics available?**

A.   Yes.

Q51. **How can an operator find out the address of the nearest authorised dealer?**

A.   By looking in the Yellow Pages for that area.

Q52. **Is there any fixed, ideal planned maintenance?**

A.   No.

Q53. **How should operators decide upon which system of planned maintenance to use?**

A.   The *Guide to Roadworthiness* (see Question 27) is exactly what it says. It gives certain suggestions, but it is for operators to decide which plan they should incorporate, depending upon the size, number and age of their fleets.

Q54. **Can any plan be recommended?**

A.   Yes. The system of planned maintenance shown in the box below has been evolved in the West Midlands and South Wales Traffic Areas over a number of years and has been shown by operators putting it into effect to be 100% satisfactory.

A SYSTEM OF PLANNED MAINTENANCE

This system consists of the following.

- A flow chart displayed on the office wall. It is recommended that the flow chart should be colour-coded and should be used to an operator's advantage. Planned inspections should be marked and dates for the servicing of vehicles, the next MOTs, the inspection (every two years) and calibration (every six years) of tachographs and all other relevant information should be shown. Retrospective information can also be included to give a complete picture.

- The use of duplicate or triplicate books by drivers for daily reporting either of defects, or of the fact that there are no defects.

- The writing of letters setting out duties for drivers. These letters should be in duplicate and signed by the drivers to confirm that they have been received, read and understood.

- Where the inspections are done in-house, letters in duplicate should be sent to the maintenance fitters.

- Vehicles should always be in A1 condition and it is recommended that vehicles should be subjected to a fresh MOT check before this simple maintenance system begins.

Q55.   Is it possible to elaborate upon the system of planned maintenance?

A.   Yes, as follows.

- There is no prescribed form of flow chart but whatever is used should be large, clear and colour-coded, even if details of only one vehicle are shown. Thus the operator, or person in charge of operations, can quickly see whether or not a vehicle is available on a specific date and time to do particular work. If it is due to be inspected on that day and time, it is unavailable.

- Inspections should be plotted on the flow chart for at least 12 months forward and at the end of each 6 month period a further 6 months should be plotted.

- Pre-arranging inspections with outside contractors is very important. If inspections are pre-booked, both the contractor and operator know that the vehicle is due for inspection on that day. Under this scheme, inspections can neither be missed nor delayed, so in the event of a vehicle inspector calling, the documentation is readily available and up to date.

- The record sheets which had defects marked on them should be kept in a separate folder, with notes relating to the remedial action taken. These sheets would then form part of the documentation which must be maintained for 15 months (see Question 62).

- The reporting books must be personal to the driver and not to the vehicle. In the road haulage industry, drivers do not tend to interchange, but they do so occasionally and it is a further safeguard to operators that if there are two drivers driving a specific vehicle on one day, then there should be two separate reports. Thus each driver should report "blind" to what the previous driver reported. Where a fleet consists of five or more vehicles, triplicate books should be used. One copy should go directly to the operator, or in the case of larger fleets, to the transport manager. The second copy should go to the foreman fitter. The third copy remains in the book, which forms part of the maintenance documentation and must be kept for at least 15 months. The reporting books should have consecutive numbering so that no sheets can be removed, showing that the record is complete.

- Letters to the drivers should be in duplicate and consist of instructions to the driver to carry out a daily cosmetic check of the vehicle, eg checking lights and reflectors (and light coupling systems, for trailers and articulated vehicles), windscreen wipers and washers, horns, tyres, steering and brakes. Instructions should also be given in the letter as to where and to whom the daily reporting sheets should be delivered. The letters should include all relevant information that is necessary for drivers to be able to comply with the law and the operator's instructions. They must be in duplicate and the copy must be signed by drivers as being received, read and understood. It is suggested that letters should be updated every six months.

Q56. How good is the prescribed system of maintenance?

A. It works if employees make it work; drivers must report defects and the fitters must repair defects. It is essential for operators to agree that any defects reported which relate to roadworthiness should be remedied immediately, before vehicles are put into further service.

Q57. If drivers have been employed for a long time, is it still necessary to provide them with letters setting out their duties?

A. Yes. A letter in duplicate should be sent to all drivers, both full and part time. Operators should ensure that the duplicates are signed as received, read and understood. It is suggested that these letters should be updated every six months.

Q58. **If only a few drivers are employed would it not be better to put a notice in the drivers' room or in the drivers' cabs?**

A.   No. The written letter, countersigned by a driver, has been proved time and time again to be the only satisfactory way to achieve the necessary direct communication between operator and driver.

Q59. **Why should drivers' reports be daily?**

A.   A defect on a vehicle can arise at any time. If weekly reporting is the norm then a defect which occurs at the beginning of the week will not be reported until the end. That defect may be one which constitutes a danger to the public and if the vehicle was stopped by Vehicle Inspectorate officers at a roadside check, it could then receive a form PG9 indicative of neglect.

Daily reporting concentrates drivers' minds and the fact that they have to hand in a piece of paper at the end of each day or journey is an *aide-mémoire* to any defect which may have occurred, however minor, earlier on the journey.

Q60. **Why should reporting books be personal to the driver? Would it not be better if they were personal to the vehicle?**

A.   No. The idea of having books personal to drivers is so that where there is more than one driver there will be a double-check on that vehicle. This prevents a second driver automatically countersigning sheets relating to a specific vehicle on the assumption that the first driver has not missed any defects.

Q61. **Why should sheets in the reporting book be consecutively numbered?**

A.   This is a satisfactory way to help operators show the Vehicle Inspectorate, on inspection, that there is a record for the vehicle for each day that it has been in service. If sheets are not consecutively numbered there is always the danger that a sheet may be mislaid or substituted at a later date. The consecutive numbering of pages thus eliminates any difficulties that may ensue.

Q62. **How long should reporting books be kept for?**

A.   They should be kept for 15 months. The carbon copies become part of the documentation but the top copies are not saved after 21 days except those that indicate defects.

Q63. **Is it advisable to have a repair book?**

A.   Yes, and by doing so accountancy fees can be saved.

Q64. **How can accountancy fees be saved by keeping a repair book?**

A.   At the end of each financial year, operators will be asked by their account-ants or financial advisors to produce a list of all the expenditure incurred upon each vehicle. This information is required so that income tax relief on the expenditure can be obtained.

A repair book should be kept either for the fleet or one for each individual vehicle. Every time there is a repair, details should be recorded of the date, the vehicle number, the repair and the cost. All that has to be done at the end of the financial year is to add up the items, photocopy the sheets and present them to the accountant. These details, if submitted correctly, could lead to a considerable saving in fees.

Q65. **What is the time period of an inspection?**

A.   There is no fixed time, but a six week interval is frequently taken as a starting point. The time period should be based on the mileage, age and general condition of the vehicle; a periodic inspection greater than six weeks for a vehicle that is constantly on the road is not recommended.

Q66. **Is it possible to say whether or not there is an average time for inspections to be carried out?**

A.   No, but it would appear that a four weekly interval forms the average period between inspections.

Q67. **If a vehicle is only used seasonally, how should the periodic times be arranged?**

A.   The vehicle should be inspected immediately before it is first put into use at the beginning of the season and thereafter it should be inspected at regular intervals during the season.

Q68. **Is it possible to conduct an MOT on a vehicle even though there is life in the current MOT certificate?**

A.   Yes.

Q69. **Which certificate is then the live one?**

A.   Both are in fact live until the date of their expiry, but it is recommended that the former certificate be documented only and that the new certificate be treated as the current one.

Q70. **Is it possible to obtain a convenient appointment at a testing station?**

A. Yes. By telephoning the testing station nearest to the operator's base, appointments which are convenient to the operator can frequently be made. It is not advisable merely to turn up at a testing station.

Q71. **What form does the test take?**

A. All items that are tested are listed and numbered in the *Heavy Goods Vehicle Inspection Manual* (see Question 30).

Q72. **With vehicles travelling over rough terrain, is it advisable to have them inspected more frequently?**

A. Yes. If vehicles travel over rough terrain or are doing particularly heavy, dirty work a more regular inspection can be beneficial. Many vehicles used on such work are inspected at seven day intervals.

Q73. **With busy operators how is it possible to have vehicles inspected without losing operational time?**

A. Most commercial garages who carry out heavy goods vehicle inspections are well aware of this problem and therefore work staggered hours, including evenings and weekends. Make sure, when choosing outside contractors, that they are capable of inspecting the vehicles at a time and date which is convenient to you and consistent with your forward planning schedules.

TRANSPORT MANAGERS

Q74. **What is a *transport manager*?**

A. A transport manager is the operational member of staff who is in charge of vehicles, their maintenance, the inspection of tachographs, the checking of weights and all matters relating to goods vehicle operations.

Q75. **Is a transport manager restricted to a certain number of vehicles?**

A. No. A transport manager can act for more than one operator.

Q76. **Does every operator have to have a transport manager?**

A. No, an "own account" restricted "O" licence holder is not required by law to have a transport manager.

Q77. **Does a Standard National or Standard International Licence operator have to have a transport manager?**

A. Yes, but he or she may be employed part time.

Q78. **How does an operator decide that a part time transport manager is what is required?**

A. The following questions should be considered when making this decision.

1. For how many (and which) hours per week will the transport manager be in your employment and engaged in the management of transport operations?

2. How many employers does the transport manager have?

3. How many vehicles does each employer have?

4. What is the distance between each of the operating centres for which the transport manager has responsibility?

5. What are the transport manager's duties — do they include being a driver and/or fitter?

6. Where does the transport manager live?

7. Is the transport manager "of good repute", ie has he or she any convictions or ever had his or her "good repute" removed by a Traffic Commissioner at a public inquiry?

Q79. **Does a transport manager have to have specific qualifications?**

A. Yes. Transport managers must either:

- have acquired what are described as "Grandfather Rights" (see also Question 101), which means that they had to prove that they had substantial involvement in the industry prior to 1974 (this method of obtaining a Certificate of Professional Conduct (CPC) without having to sit an examination is no longer available), or

- obtain a Certificate of Professional Competence (CPC).

Q80. **What is a Certificate of Professional Competence?**

A. This certificate is obtained by passing an examination which is set by the Royal Society of Arts (RSA).

Q81. **How can a prospective transport manager find out about the examinations?**

A. By checking details in Croner's *Road Transport Operation* or by contacting the RSA Examinations Board, Westwood Way, Westwood Business Park, Coventry, CV4 8HS (tel: 01203 470033).

Q82. **Can a transport manager lose his or her "good repute"?**

A. Yes. The Traffic Commissioners can remove it.

Q83. **Can a transport manager appeal against the decision to remove his or her "good repute"?**

A. At present, no, but the regulations are being altered to allow a transport manager who is not an operator to appeal. Of course, if the transport manager is also the operator, then in his or her position as operator, an appeal to the Transport Tribunal can be made against any decision of the Traffic Commissioner.

Q84. **What is the Transport Tribunal and where is it based?**

A. The Transport Tribunal is the appellant court dealing with all appeals resulting from Traffic Commissioners' decisions. The Transport Tribunal can be contacted at 48–49 Chancery Lane, London WC1A 1JR (tel: 0171-936 7494).

DRIVER LICENSING

Q85. **What is the main legislation governing driver licensing for large goods vehicle (LGV) and passenger carrying vehicle (PCV) drivers?**

A.
- **Road Traffic Act 1988**
- **Road Traffic (Driver Licensing and Information Systems) Act 1989**
- **Motor Vehicles (Driving Licences) (Large Goods and Passenger Carrying Vehicles) Regulations 1990** (SI 1990 No. 2612)
- **Motor Vehicles (Driving Licences) (Heavy Goods and Public Service Vehicles) Regulations 1990** (SI 1990 No. 2611).

The two sets of regulations above came into force on 1 April 1991.

Q86. **What was the major change so far as the licence document itself is concerned?**

A. All driving entitlement is now shown on one driving licence. The Secretary of State for Transport has been the licensing authority since 1 April 1991. HGV licenses have now been replaced by LGV entitlement. PSV (public service vehicle) licences continue to be valid until they expire.

Q87. **How do the previous groups equate with the new categories of entitlement?**

DETAILS OF THE NEW CATEGORIES

Type of vehicle	New category	Old group or class
Motor car or light goods vehicle up to 9 seats/3500kg but fitted with automatic transmission	B (fitted with automatic transmission)	B
Motor tricycle or three-wheeled car or van up to 500kg unladen mass (not shown on a full category B licence: category B also covers three-wheeled vehicles up to nine seats/3500kg)	B1	C
Invalid carriage	B1 (limited to invalid carriages)	J

Goods vehicle with maximum authorised mass exceeding 3500kg but not more than 7500kg	C1	A
Goods vehicle as above plus trailer of more than 750kg with a combined weight of not more than 8250kg	C1 + E	A
Large goods vehicle with maximum authorised mass more than 3500kg	C	HGV class 2 or 3
Large goods vehicle with drawbar trailer of more than 750kg	C + E (limited to drawbar trailers)	HGV class 2 or 3
Articulated large goods vehicle	C + E	HGV class 1
Passenger carrying vehicle (9–16 passenger seats) not used for hire or reward	D1 (not for hire or reward)	A
Category D1 with trailer of more than 750kg	D1 + E	A
Passenger carrying vehicle (9–16 passenger seats) — unrestricted use	D (limited to 16 passenger seats)	New category
Passenger carrying vehicle (more than eight passenger seats and not longer than 5.5m)	D (not more than 5.5m long)	PSV class 4
Large passenger carrying vehicle (more than eight passenger seats and longer than 5.5m)	D	PSV class 3
Articulated bus and bus towing trailer of more than 750kg	D + E	PSV class 1 or 2
Category E is not used by itself but when another category is shown against it this gives entitlement to draw a trailer of more than 750kg		
Agricultural tractor	F	F
Road roller	G	G

Tracked vehicle	H	H
Mowing machine/Pedestrian controlled vehicle	K	K
Electrically propelled vehicle	L	L
Duty exempt vehicle	N	N
Moped	P	E
Motorbicycle (with or without sidecar) or scooter	A	D
Motor car or light goods vehicle up to 9 seats/3500kg	B	A

Source: DVLA

Q88.   Do the new categories show any specific change and, if so, what?

A.   The new categories which have replaced the groups are the same as those used throughout the EU. Classification no longer depends on the number of axles, but relates more closely to the type of vehicle. **Note**: some of the category definitions are due to change on 1 July 1996, following implementation of the second EC Directive on driver licensing.

Q89.   Now that there is one licence, does this mean that every LGV/PCV driver has to have a new style one?

A.   Yes, eventually. There are some licences issued by Traffic Area Offices (TAOs) which are still current, but when they expire they will be renewed by the DVLA. TAOs ceased to issue HGV licences on 31 March 1994, and will cease to issue PSV licences on 31 March 1996.

Q90.   Were there any changes in the driving test following 1 April 1991?

A.   There are differences in the minimum test vehicle requirements as per regulation 18 of the **Motor Vehicles (Driving Licences) (Large Goods and Passenger Carrying Vehicles) Regulations 1990** (SI 1990 No. 2612). Refer to the Driving Standards Agency (DSA) for the actual test content.

Q91.   What are the addresses and telephone numbers of the DSA to whom an application should be made for an LGV or PCV driving test?

A.   These contact details are shown in the box overleaf.

DSA ADDRESSES AND TEST CENTRES

DSA REGION	ADDRESS	TEST CENTRES
North Eastern	Westgate House, Westgate Road, Newcastle upon Tyne NE1 1TW Tel: 0191-261 0031	Berwick, Beverley, Darlington, Grimsby, Keighley, Leeds, Newcastle, Sheffield, Walton (York)
North Western	Portcullis House, Seymour Grove, Stretford, Manchester M16 0NE Tel: 0161-876 4474	Bredbury (Manchester), Caernarfon, Carlisle, Heywood (Manchester), Kirkham (Preston) Simonswood (Liverpool), Upton (Birkenhead), Wrexham
West Midlands	Cumberland House, 200 Broad Street, Birmingham B15 1TD Tel: 0121-631 3300 (ext. 611/612)	Featherstone (Wolverhampton), Garrets Green (Birmingham), Shrewsbury, Swynnerton (Stoke on Trent)
Eastern (Nottingham)	Birkbeck House, 14–16 Trinity Square, Nottingham NG1 4BA Tel: 0115 955 7600	Alvaston (Derby), Leicester, Watnall (Nottingham), Weedon
Eastern (Cambridge)	Terrington House, 13–15 Hills Road, Cambridge CB2 1NP Tel: 01223 321396/7/8	Chelmsford, Ipswich, Leighton Buzzard, Norwich, Peterborough, Waterbeach (Cambridge)
South Wales	Caradog House, 1–6 St. Andrew's Place, Cardiff, South Glamorgan CF1 3PW Tel: 01222 225186/7/8	Llantrisant, Neath, Pontypool, Withybush (Haverfordwest)
Western	The Gaunt's House, Denmark Street, Bristol BS1 5DR Tel: 0117 929 7221	Bristol, Camborne, Chiseldon (Swindon), Exeter, Gloucester, Plymouth, Poole, Taunton

South Eastern	Ivy House, 3 Ivy Terrace, Eastbourne BN21 4QT Tel: 01323 646124/5	Canterbury, Culham, Gillingham, Hastings, Isle of Wight, Lancing, Reading, Southampton
Metropolitan	PO Box 2224, Charles House, 375 Kensington High Street, London W14 8TY Tel: 0171-605 0399/ 0400/0401/0408	Croydon, Enfield, Guildford, Purfleet, Yeading
Scotland	83 Princes Street, Edinburgh EH2 2ER Tel: 0131-225 7164	Aberdeen, Bishopbriggs (Glasgow), Connel, Dumfries, Elgin (Keith), Galashiels, Inverness, Kilmarnock, Kirkwall, Lerwick, Livingston (Edinburgh), Machrihanish (Kintyre), Perth, Port Ellen (Islay), Stornoway, Wick

Q92. **Is an LGV entitlement the same as a previous HGV licence?**

A. Yes, in broad terms. Category C gives entitlement to drive all rigid goods vehicles, and category C + E gives entitlement to drive articulated goods vehicles and goods vehicles with large trailers.

Q93. **Is a PCV entitlement the same as an old style PSV licence?**

A. Yes, in broad terms. Category D gives entitlement to drive all buses, and category D + E gives entitlement to drive articulated buses and buses with large trailers.

Q94. **When heavy goods vehicle (HGV) licences became described as large goods vehicle (LGV) entitlements, and public service vehicle (PSV) licences became passenger carrying vehicle (PCV) entitlements, why is it that LGVs are still referred to as HGVs and PCVs as PSVs?**

A. Purely as a matter of habit.

Q95. **Will there be a later change in this nomenclature?**

A. No change is required; the new titles have already been introduced by the 1990 regulations.

Q96. **At what age can the holder of an ordinary driving licence apply for an LGV or a PCV entitlement?**

A.   The minimum age for driving a *goods vehicle* over 7500kg maximum authorised mass is normally 21. However, there are two exceptions as follows.

- For full or part-time members of the armed forces when driving official vehicles for Crown purposes, the minimum age is 17.
- For members of the Young Large Goods Vehicle Trainee Drivers' Scheme, the minimum age is 18.

The minimum age for driving a *passenger carrying vehicle* with more than eight passenger seats is also normally 21. However, it is 18 when the following conditions apply.

- While learning to drive a category D vehicle under the supervision of the holder of a category D entitlement.
- While taking the category D test.
- When the vehicle is not engaged in the carriage of passengers and the driver has category D entitlement.
- When the vehicle is engaged in the carriage of passengers and is being driven under a PCV Operator's Licence, a small or large bus permit or a community bus permit by the holder of category D entitlement driving a vehicle which is either:
  - on a regular service where the route length is no more than 50km (31 miles), or
  - constructed or adapted to carry no more than 16 passengers and the journey is within the UK only.

Q97. **When driving under a provisional LGV licence, does the driver have to carry "L" plates?**

A.   Yes, on the front and back of the vehicle.

Q98. **What are the entitlements of minibus drivers following 1 April 1991?**

A.   Holders of a full category B (motor car) entitlement continue to be able to drive minibuses with up to 16 passenger seats not for hire or reward (ie category D1). They may also drive larger buses under the Large Bus and Community Bus Permit Schemes.

Q99. **Can the holder of a full ordinary driving licence obtained prior to 1 April 1991 drive a bus which is not in service?**

A. It depends on whether the vehicle is being used on a road. If it is, normal licence requirements apply.

Q100. **Do the new regulations affect drivers who were described as "movers" prior to 1 April 1991 and fitters who drive vehicles without passengers in order to test the vehicle?**

A. Drivers will need to have either category D, D with a category restriction not for hire or reward or C entitlement.

Q101. **What are "Grandfather Rights"?**

A. This is an administrative scheme which enables licence holders to continue to retain entitlement which a change in legislation may remove, but which may unfairly disadvantage certain individuals.

Q102. **Can "Grandfather Rights" be acquired for any new entitlement?**

A. No. They only apply in specific circumstances, eg that for large buses expired on 30 September 1992.

Q103. **What is the position with regard to a licence required to drive a goods vehicle which is within the Operator's Licence and is over 3500kg but under 7500kg?**

A. Holders of full category B entitlement may drive a medium sized goods vehicle of between 3500kg and 7500kg provided they are 18 or over.

Q104. **Prior to 1 April 1991 the issuing and control of vocational licences was dealt with by the Traffic Area offices. What is their function now in relation to the issuing and conduct of vocational driver entitlement?**

A. They issue duplicates of existing old style PSV licences and have full responsibility for conduct — as before.

Q105. **As the Traffic Commissioners are agents for the Secretary of State for Transport, who is the licensing authority now?**

A. The DVLA is the licensing authority on behalf of the Secretary of State.

Q106. **Is the Secretary of State for Transport bound by any decision that the Traffic Commissioner may make?**

A. Any decision the Traffic Commissioners make with regard to conduct is binding on the DVLA.

Q107. **What is the position with regard to medical examinations and further periodic medical examinations required in order to maintain an LGV or PCV entitlement?**

A. A medical report is required with the first licence application and at every subsequent renewal where the driver is over 45 years of age. Any illness or disability which develops or worsens in the interim must, by law, be reported to the DVLA, unless it is likely to get better within three months.

Q108. **What fees are payable in respect of the medical examination?**

A. The applicant has to pay the GP a fee for the completion of the medical report — the fee recommended by the BMA is £46.50. Any subsequent enquiries or investigations by the DVLA's Medical Advisor are free of charge.

Q109. **Can an applicant be refused a vocational entitlement on medical grounds?**

A. Yes.

Q110. **What medical grounds should drivers be aware of which could bar them from holding LGV or PCV licences?**

A. In brief, the following:
- a liability to epileptic seizures
- defective eyesight, including monocular vision
- insulin-treated diabetes.

Q111. **Does a driver have the right to appeal against the medical decision?**

A. Yes, but not against medical standards prescribed by legislation.

Q112. **If the appeal is unsuccessful in the magistrates' court, can the DVLA apply for costs against the driver who has appealed?**

A. Yes.

Q113. **Does the driver have to pay for his or her own medical evidence, which he or she brings to refute the evidence given by the DVLA doctors?**

A. Yes.

Q114. **Can the driver apply for Legal Aid in respect of such an application?**

A. No.

Q115. **Are there any other methods of obtaining financial assistance?**

A. Yes. The Transport and General Workers Union, and other unions who have members who hold vocational entitlements, do their utmost to protect the interests of their members. However, this does not necessarily mean that they would pay for the services of a solicitor or barrister, but advice is always available from the union concerned.

Q116. **What is the medical report form D4 (previously DTp 20003) and is the driver entitled to see it?**

A. Form D4 is a questionnaire completed by a GP following an examination to determine an applicant's fitness to drive LGVs and PCVs. One must be submitted with every first application and subsequent renewals for LGV and PCV entitlements. The applicant has to sign the completed form.

Q117. **What special requirements are there for a medical examination and submission of a medical form D4?**

A. Higher medical standards apply to drivers of LGVs and PCVs. The Secretary of State must be satisfied that anyone applying for a vocational entitlement has no condition likely to impair driving ability. Every application for a vocational licence (provisional or full) has to be accompanied by a Medical Examination Report.

Q118. **Where should a driver direct any enquiries relating to medical matters?**

A. To the Drivers Medical Unit, DVLC, Swansea SA99 1TU. Tel: 01792 783705.

Q119. **Is it always necessary to give a date of birth and postcode when communicating with the DVLA?**

A. To access a driver's record, the DVLA needs the driver's full name and date of birth, or the driver number. The postcode is not essential for tracing a record, but it does help for despatch and delivery purposes.

Q120. **Is it true that each driving licence number is based on the holder's date of birth?**

A. A driver number (the identification for every driver) is made up of the driver's name and date of birth, together with other characters to make it unique (see box overleaf).

EXAMPLE OF HOW A DRIVING LICENCE NUMBER IS MADE UP BY THE DVLA

Driver Number

A	B	C	D	E
DRIVE	512185	YT	9	ME

A = First five characters of the driver's surname. If the surname has fewer than five characters, letters will be replaced by the figure 9 (eg MAN99).

B = First and last characters are the year of birth (in this case 1955). Second two characters are the month of birth (in this case the 12th month). (**Note**: in the case of a female licence holder, 50 is added to the month of birth, so the second digit in this box will always be a 5 or a 6.) The fourth and fifth digits are the day of the month (in this case the 18th).

C = The first two initials of the forenames; where there is only one initial, the second letter will be replaced by the figure 9.

D = A tie-breaker digit in the event of identical computer personal details.

E = Computer check digits to prevent forgery.

Q121. Which department at the DVLA should drivers communicate with about their driving licences?

A. Enquiries should be addressed to the Customer Enquiries Unit, DVLC, Swansea SA6 7JL. Tel: 01792 772151 between 0815–1630 Monday to Friday. For Minicom Users only — Tel: 01792 782756.

Q122. If a licence is lost, can a duplicate be obtained with the same entitlements upon it?

A. Yes. Duplicate PSV licences are available from Traffic Area Offices or duplicate, fully harmonised licences are available from the DVLA.

Q123. What is the cost of a duplicate licence?

A. £6.

Q124. What are the costs of an ordinary licence and provisional LGV and PCV licences?

A. As at August 1995 the costs of licences are as set out in the table opposite.

COSTS OF LICENCES

First provisional licence	£21
First full licence after passing a test	free
Duplicate licence	£6
Exchange licence (eg claiming additional entitlement, removing endorsements, adding or removing provisional motorcycle entitlement to category A)	£6
Issuing new licence after a period of disqualification	£12 and £21 (for provisional LGV/PCV)
Renewal	£6/£21 (LGV/PCV)
First full GB licence	£21
Exchanging a Northern Ireland full licence for a GB licence	£6
Medical renewal	free

Note: These fees apply to all categories unless stated otherwise.

Q125.   What are the costs of the various driving tests?

A.   As at August 1995 driving tests costs are as set out below.

COSTS OF DRIVING TESTS

Car	Weekday	£28.50
	After 1630	£38.50
	Saturday	£38.50
Motorcycle	Weekday	£36.00
	After 1630	£47.50
	Saturday	£47.50
LGV/PCV	Weekday	£62.00
	Saturday	£80.00

For extended tests following disqualification:		
Car	Weekday	£57.00
	Saturday	£77.50
Motorcycle	Weekday	£72.00
	Saturday	£95.00
CBT Certificates		£5.00

Q126. **Are test fees payable in advance?**

A. Yes.

Q127. **When the test is taken will the examiner require to see the provisional licence?**

A. Not necessarily, but it is the best form of identification to use. The licence holder should sign it as soon as it is obtained.

Q128. **Are examiners insured for taking tests, or do they require to see the insurance of the vehicle in which a driver is taking a test?**

A. Examiners are insured.

Q129. **If a motoring offence is committed and penalty points are endorsed on the counterpart to the harmonised licence, does this affect the lorry/bus entitlement?**

A. The DVLA will inform the Traffic Commissioners of any LGV/PCV licence holder who receives a disqualification of 56 days or more, or a drink or drugs-related endorsement. The Traffic Commissioners will consider the driver's conduct and its impact on his or her entitlement. However, in the case of a participant in the Young Drivers' Scheme (ie a driver under 21 years of age) or a young military driver, there is a statutory requirement that if more than three penalty points are accrued, the LGV entitlement should be revoked at least until the holder's 21st birthday. This is normally dealt with by the DVLA.

Q130. **If there are penalty points endorsed on a car licence and an application is made for a provisional PCV or LGV, is this looked at by the Traffic Commissioners?**

A. Yes. The criteria for referral to the Traffic Commissioners by the DVLA on receipt of a new application for LGV or PCV entitlement are:

- those who have accumulated nine or more penalty points during the last two years
- those who have been disqualified for any length of time/reason during the last four years
- those who have been convicted of a drink or drugs-related offence during the last four years.

The applicant may be seen personally by a Traffic Commissioner and warned as to his or her future conduct, and further advised about the higher standards required from LGV/PCV drivers. Furthermore, the Traffic Commissioner may decide that because of prior driving conduct the applicant should wait for a period before he or she can obtain a provisional licence.

Q131. **How long does a provisional licence last and can a new application be made for renewal after its expiry?**

A.   The provisional car licence lasts until the holder's 70th birthday and thereafter is renewable every three years at most.

The LGV/PCV provisional entitlement lasts until the holder's 45th birthday and is then renewable every five years until the holder is 65, after which it is renewable annually.

Q132. **Is there any period in which a test must be taken after the grant of a provisional licence?**

A.   There is no limit to the period in which the test must be taken.

Q133. **If an LGV driver is convicted of motoring offences which are not endorsable, does that affect his or her holding of the lorry/bus entitlement?**

A.   Any court convictions must be declared to the Traffic Commissioner who will decide what action, if any, should be taken with regard to the driver's lorry/bus entitlement.

Q134. **If a PCV driver commits offences, other than motoring offences, can they in any way affect his or her holding of a PCV licence?**

A.   See the answer to Question 133 above.

Q135. **What is the difference between offences committed by LGV and PCV drivers?**

A.   LGV drivers are disciplined for committing motoring offences, whereas with PCV drivers their conduct as a whole is examined, including, for example, any previous sex offences. This is because PCV drivers are responsible for the safe transportation of passengers and the Secretary of State has to be assured that the drivers are (in relation to their conduct) fit to hold PCV entitlement as well as presenting no danger to other motorists.

Q136. **When offences are committed which can affect the vocational entitlement, how is the driver notified of further action?**

A. The Traffic Commissioner will write to the driver.

Q137. **While the magistrates' courts or the Crown Court deal with motoring offences why do they not also deal with offences relating to vocational entitlement?**

A. Because they have no power to do so. The disciplining of vocational drivers is the responsibility of the Secretary of State, represented by the Traffic Commissioners who act as his or her agents.

Q138. **If the offender is a HGV or PSV driver and is still in possession of the old licence, does that licence have to be handed to the magistrates at the time of endorsement or disqualification?**

A. No, but the offence must be notified to the Traffic Area Office.

Q139. **If the entitlement is on the new harmonised licence, rather than the old style licence, does this still mean that the magistrates can only take action against the car entitlement?**

A. No. Magistrates have powers to deal with the entitlement to drive in its entirety. They cannot apply a disqualification only to car, or lorry or bus entitlement.

Q140. **If a driver is disqualified from holding a car driving licence as a result of committing an offence, does this automatically revoke any lorry or bus entitlement?**

A. Yes, because any lorry or bus entitlement is dependent on the holder of the licence holding a full car licence.

Q141. **Can an applicant apply for a provisional car licence and at the same time for a provisional LGV/PCV licence?**

A. No, because a provisional LGV and PCV can only be considered when the applicant is already the holder of a full car licence. The only exception to this is when the drivers are members of the armed forces.

Q142. **Can action be taken by a Traffic Commissioner when any penalty points are endorsed upon the car licence?**

A. Yes, if the driver also has lorry or bus entitlement. However, the Traffic Commissioner cannot, for example, extend a period of disqualification from driving cars.

Q143. **Is it true that any endorsable offence also carries with it a possible disqualification?**

A. Normally, where penalty points incurred on a minor offence bring the overall total accrued to 12 or more points in three years, a disqualification under the "totting up" procedure would apply. However, the court will use its discretion to impose the penalty which it thinks is appropriate.

Q144. **Under the "totting up" procedure in magistrates' courts, when a driver has 12 or more penalty points, is the driver automatically disqualified?**

A. Not necessarily. In most cases, the driver will be disqualified, but in cases of *exceptional* hardship the magistrates will exercise discretion.

Q145. **Does the Secretary of State have a separate points system or a number of points when action is taken against the LGV/PCV entitlement?**

A. No, there is no separate system. The Secretary of State relies entirely on penalty points endorsed upon the car driving licence and convictions (or non-endorsable offences) relating to goods vehicles. However, the commission of an offence by the holder of a LGV/PCV entitlement may well mean that the holder will receive a warning from the Traffic Commissioner acting as agent for the Secretary of State as to his or her future conduct, or the holder may be called before the Traffic Commissioner with a view to the LGV/PCV entitlement being suspended or revoked.

Q146. **What form does such a warning take?**

A. It takes the form of a letter to say that if any further offences are committed in any vehicle at any time, which result in any further penalty points being endorsed upon the licence, then the licence is in jeopardy.

Q147. **Does this mean that if an offence is committed in a private motor car it can affect the LGV/PCV entitlement?**

A. Yes. Remember that the LGV/PCV entitlement when first granted will depend very much upon the penalty points already existing.

Q148. **Can the LGV/PCV entitlement be revoked whilst the car entitlement remains and what vehicles, if any, can then be driven?**

A.   LGV/PCV entitlement can be revoked and this does not affect the validity of the car licence. The holder of a car licence with a revoked LGV/PCV entitlement can drive goods vehicles up to 7500kg, plated weight, and passenger carrying vehicles not exceeding 16 passenger seats (not used for hire or reward).

Q149. **Is it true that an LGV entitlement may, in the future, be required for driving a vehicle in excess of 3500kg?**

A.   Yes. From 1 July 1996 new licence holders wishing to drive category C1 (3500kg to 7500kg) vehicles will need to take a separate test.

Q150. **If the LGV/PCV entitlement is revoked, when can a new application be made?**

A.   A driver may apply at any time to the Secretary of State for a new LGV/PCV entitlement, unless the holder has been disqualified for a period of time from holding such entitlement.

Q151. **Is revocation generally accompanied by a period of disqualification?**

A.   Yes.

Q152. **Is there any set period of disqualification in relation to any offences?**

A.   No. It is a matter entirely for the discretion of the Traffic Commissioner. However, if the holder of the LGV/PCV entitlement has previously been warned as to his or her future conduct it is quite likely that the LGV/PCV entitlement will either be suspended for a period of time or revoked with a period of disqualification restricting the individual from obtaining or holding such an entitlement.

Q153. **If a Traffic Commissioner wishes to take action against a driver's LGV/PCV entitlement does he or she first of all notify the holder of this fact?**

A.   Yes, in the interests of natural justice. If a Traffic Commissioner intends to suspend or revoke the entitlement and for some reason it has not been possible for him or her to see the holder, a letter will be sent informing the holder of the intention to suspend or revoke and inviting the holder to put forward any reasons for further consideration.

Q154. **Do Traffic Commissioners have power to call the holder of a LGV/PCV entitlement before them?**

A. The general form is to invite the holder of the LGV/PCV entitlement to come to either a traffic court, which is referred to as a public inquiry, or to see the Traffic Commissioner, personally, in chambers. Sometimes it is not possible, eg because of distance, for a driver to travel to the Traffic Commissioner's office and, in those circumstances, the matter is dealt with by letter.

Q155. **After a Traffic Commissioner has made a decision, what right of appeal is there?**

A. The holder of a LGV/PCV entitlement who is dissatisfied with any decision made by the Traffic Commissioner may appeal to the local magistrates' court within six months of the decision of the Traffic Commissioner, or, in Scotland, to a sheriff's court within three weeks.

Q156. **Is any decision the Traffic Commissioner makes stayed, pending the hearing of an appeal?**

A. No. If a driver is suspended or disqualified and wishes to appeal against that decision, he or she can request that a stay of execution is granted pending the conclusion of the appeal. This request may be made either to the Traffic Commissioner who made the order or to the magistrates before whom the appeal will be made.

Q157. **Is any stay application bound to succeed?**

A. No, it will depend very much on the facts. Traffic Commissioners tend to impose very short periods of suspension, more as a mark of warning. Thus an appeal would have to be made very quickly in such circumstances.

Q158. **Do appeals generally succeed?**

A. Every case is determined upon its facts. The majority of appeals have been rejected by magistrates in favour of the original decisions of the Traffic Commissioners.

Q159. **What are the chances of success in a medical case appeal before magistrates?**

A. This will depend very much upon the evidence and, in particular, the medical evidence of the case.

Q160. **Can an appellant insist upon a medically qualified person sitting on the bench of magistrates presiding over the case?**

A. No. Magistrates' court rotas are determined well in advance. It is not unknown for a magistrates' clerk to rearrange a court list so that a magistrate who is an expert in a certain field, eg a member of the medical profession, can be available to sit with fellow magistrates on a particular case.

Q161. **Does a magistrates' court have the power to award costs?**

A. Yes, and costs in appeals can be considerable. Before any appeal is undertaken the driver who is aggrieved is recommended to seek legal advice. Even if the driver were to represent himself or herself and lost the case, he or she could still be ordered to pay the costs of the defendant; in this case the Secretary of State for Transport.

Q162. **Can a driver with a LGV/PCV entitlement request a meeting with the Traffic Commissioner to discuss his or her penalty points?**

A. Yes. A Traffic Commissioner will always see a driver in private should a request be made. When drivers are seen *en masse*, a proviso is always made that should any driver wish to see the Traffic Commissioner personally and privately, rather than giving evidence in public, a private meeting is always possible if the driver waits until the end of the list.

Q163. **What form do hearings before the Traffic Commissioners take?**

A. Some Traffic Commissioners see drivers *en masse* to explain their powers and the reasons why it is necessary for LGV/PCV entitlement holders to be good drivers at all times. These talks normally last approximately 15 minutes. Each driver is then seen in front of the other drivers to give his or her side of the case. A driver may be represented by a lawyer, a friend, a trade union official or anyone else that he or she wishes to be supported by. The driver may call witnesses if desired.

Other Traffic Commissioners will deal with the matter by correspondence.

Q164. **If the driver is seen privately by the Traffic Commissioner is there any variation in the procedures?**

A. Basically, no, although the proceedings tend to be more informal. The same rights of audience and calling of witnesses apply.

Q165. **If a suspension is ordered can it be said to commence at a future date or must it be immediate?**

A. It can start on any date that the Traffic Commissioner decides.

Q166. **What sort of record is taken at public inquiries?**

A. Generally, the proceedings are recorded and a transcript of each driver's application or disciplinary proceedings is available and is certainly produced should there be an appeal.

Q167. **Is the oath given at these public inquiries?**

A. No.

Q168. **What applications are heard by a Traffic Commissioner?**

A. Applications for a provisional LGV or PCV licence from a driver who has eight or more penalty points for offences committed prior to the date of the application. Applicants may be represented, call witnesses, or produce letters of support and references as to their good work and driving.

Q169. **What is the position of a driver who has previously been disqualified for a drink-related offence?**

A. Generally speaking, a period of rehabilitation will have to expire, during which time the driver can drive vehicles not exceeding 7500kg or passenger carrying vehicles with no more than 16 passenger seats, not used for hire or reward. This period is generally around three months, but it is a matter entirely for the discretion of the Traffic Commissioner. For example, if a driver is disqualified for a period of 12 months it is generally unlikely, unless there are exceptional circumstances, that his or her LGV/PCV entitlement would be returned until 15 months have expired.

Q170. **When a driver has been disqualified and has LGV/PCV entitlement, how can he or she restore that LGV/PCV entitlement when his or her car licence is returned?**

A. The driver must apply to the DVLA for the restoration of the LGV/PCV entitlement and that application will be determined by the Traffic Commissioner. Generally, Traffic Commissioners will see applicants who have been disqualified, although sometimes when the disqualification has been for a short period of time, the restoration may be made after the application with a warning being given as to future conduct. This again depends upon the circumstances and is at the discretion of the Traffic Commissioner.

Q171. **Are there any set rules as to the restoration of the LGV/PCV entitlement?**

A.   No. Once a car licence has been returned, indicating that the period of disqualification has been served, the holder of that licence is entitled to apply to the DVLA for the return of the additional LGV or PCV entitlement. It is then up to the Traffic Commissioner to decide when it should be returned, if at all, and if it is a new application for an entitlement, when that entitlement should come into effect, if at all.

Q172. **What happens if a driver has been disqualified more than once for drink-driving related convictions?**

A.   As a persistent offender, the driver will have been disqualified for a period of at least three years by the magistrates and the DVLA will look into his or her drinking history before returning the car licence. The Traffic Commissioner will look very carefully at drivers who have more than one drink-driving related conviction. In addition to the medical form which the DVLA will require, the Traffic Commissioner may also require the driver to produce a certificate from his or her personal doctor to the effect that the driver does not have any alcohol-related diseases or problems.

Q173. **To whom is the Traffic Commissioner responsible in making such decisions?**

A.   To the driver, the public at large and the Secretary of State for Transport.

Q174. **Can members of the public report bad driving incidents?**

A.   Yes. These reports can either be made to the police or to the Traffic Commissioner.

Q175. **What action will be taken?**

A.   This will depend entirely on the evidence. Any uncorroborated statement with regard to bad driving will be unlikely to reach the courts.

Q176. **If just the registered number of the vehicle is given, is that sufficient to trace it?**

A.   Yes. The DVLA records can trace the person in whose name the vehicle is registered from its computer records and this information can be given to the police and the Traffic Commissioners.

Q177. **Can complaints against drivers be made anonymously?**

A.   Yes, but it would be unlikely for any court proceedings to follow from an anonymous statement. If the circumstances warranted it and there was more than one anonymous complaint, there is no doubt that the Traffic Commissioner may well consider writing a warning letter to the offending driver.

Q178. What powers does the Traffic Commissioner in fact have over an LGV/PCV driver's entitlement?

A. The Traffic Commissioner may:

- take no action at all
- revoke the licence, thus enabling a new application to be made
- revoke and disqualify the holder from holding or obtaining a LGV/PCV entitlement for a fixed period of time, or
- suspend the entitlement for a period.

Finally, if the Traffic Commissioner considers that a warning as to the future conduct of that driver will suffice, this will be given.

Q179. Are warnings taken into consideration if further offences are committed?

A. Yes, and suspension may well follow a further offence.

Q180. How long does the LGV/PCV entitlement last?

A. Full and provisional vocational entitlement expires on the holder's 45th birthday. It then has to be renewed every five years until the holder is 65 and every year thereafter.

Q181. If a licence is revoked or suspended, is any part of the fee returned?

A. No.

Q182. How long do endorsements stay on an ordinary driving licence?

A. The length of time endorsements must remain on a licence varies according to the seriousness of the offence and also whether or not disqualification is involved.

The endorsement stays on the licence for 11 years from the date of conviction for drink and drug-related driving offences, but it is only effective, as far as magistrates are concerned, for 10 years from the date of conviction. The best example of this is that if a further drink-drive offence is committed during a period of 10 years from the date of the first conviction the offender will automatically be disqualified upon conviction for a minimum period of three years. If, however, the offence occurs in the period after 10 years from the date of the first conviction, even though it is in the period of the 11th year and the endorsements are still on the licence, it will be treated by the magistrates as a first offence.

Other offences for which the endorsement remains for a period of 11 years are:

- causing death by careless driving whilst under the influence of drink or drugs
- causing death by careless driving, then failing to provide a specimen for analysis.

The endorsement stays on a driving licence for four years from the date of conviction, but is only recognised by the magistrates for a period of three years for offences of reckless or dangerous driving and all other offences resulting in disqualification.

In all other cases, endorsements remain on the licence for four years from the date of the offence.

Q183. **If a driver sends an application to the DVLA to renew an existing licence, is he or she covered to drive?**

A. Under s.88(1) of the **Road Traffic Act 1988**, provided the DVLA have received a valid application and the applicant is not disqualified from driving, has not been (and would not be) refused a licence on medical grounds and keeps to the conditions that apply to that licence, then the applicant is allowed to drive whilst waiting for the licence to arrive through the post.

Q184. **What should a driver do on changing his or her name or address?**

A. The driver should complete the changes section on the back of the driving licence with the new details and send it to the DVLA. A replacement licence, in these cases, will be sent free of charge.

Q185. **When a licence holder dies, what action should the next of kin take?**

A. The licence should be returned to the DVLA.

Q186. **Are any special licences required for the driving of historic goods vehicles?**

A. No. A historic goods vehicle may be driven by the holder of category B entitlement, provided it is unladen and not drawing a laden trailer.

Q187. **How is a historic goods vehicle defined for driving licence purposes?**

A. A historic goods vehicle is defined as a vehicle manufactured before 1 January 1960.

The assistance of the Driver and Vehicle Licensing Agency (DVLA) in the preparation of this section is gratefully acknowledged.

Q188. **What is the legislation governing the taxation of vehicles?**

A. The **Vehicles Excise and Registration Act 1994**, which came into effect on 1 December 1994.

Q189. **How do I apply for a tax disc?**

A. By handing the MOT certificate, vehicle registration document and, for LGVs, certificate of compliance to the taxation office and completing the relevant form.

Q190. **Does a tax disc in the window mean that the vehicle has been taxed?**

A. If the disc is the correct one for that vehicle and the expiry date has not been passed, it may be assumed that the tax has been paid. A tax disc is nothing more than a receipt for money.

Q191. **When the tax expires do I have any leeway period in which to obtain the new tax?**

A. As a matter of law, the tax must be paid by the due date. As a matter of practice, the authorities are unlikely to take any action during the first 14 days after expiry.

Q192. **My vehicle is taxed at the concessionary rate for 38,000kg on six axles. Can I pull a two-axle trailer with it?**

A. Only if the two-axle trailer is empty. You may not carry a load of any sort unless the trailer has three axles, subject to s.16 (paragraphs 6 and 7) of the **Vehicles Excise and Registration Act 1994**, which permits the use of a single axle trailer in certain circumstances.

Q193. **My vehicle is taxed at 38,000kg on six axles. Can I "lift" an axle?**

A. Yes. The taxation category depends on the number of axles: the fact that an axle is not, for the moment, touching the road is immaterial.

Q194. **I have recently taxed a vehicle, but it will now be off the road for some time. Can I use the disc on another vehicle?**

A. No. This is one of the most common reasons for a prosecution. It is also a very serious offence amounting to the fraudulent use of a tax disc. Discs are allocated to a specific vehicle and any attempt to use a disc on another vehicle is an offence.

Q195. **My tax has run out and I will not be able to tax the vehicle for a few days. What should I do with the existing disc?**

A.   You can either leave it in the windscreen or remove it. You can put a note in the disc holder confirming that the tax has been applied for, but only if that is a true statement.

Q196. **Does the receipt of the money for the tax confirm the type of vehicle that has been taxed?**

A.   No. Whilst the taxation authorities may have a view as to the correct rate of duty, ultimately it will always be for the courts to decide what category of vehicle is involved.

Q197. **Are any vehicles exempt from tax?**

A.   Section 5 of the **Vehicles Excise and Registration Act 1994** exempts those vehicles listed in schedule 2 to the act. The list is quite extensive. If any doubt arises, the local taxation office should be consulted.

Q198. **I have a high pressure water-jetting vehicle and I understand that it is not a goods vehicle. Is this correct?**

A.   The categories of duty for goods vehicles are set out in schedule 1 of the **Vehicles Excise and Registration Act 1994**. The schedule is split into a number of parts, each part dealing with different categories of vehicle. This is a substantial change from the earlier Act. Part VIII, paragraph 14 covers the taxation of vehicles which are effectively mobile machines. This provides that such vehicles will be taxed as if the machine is effectively "burden". There is no direct indication as to how the weight should be obtained, although it remains clear that these vehicles are not "goods vehicles".

The **Finance Act 1995** suggests that such vehicles will be taxed at the "revenue weight".

Q199. **What is "revenue weight"?**

A.   This is a system of ascertaining the weight of vehicles which are not subject to plating and testing (certain vehicles are, for example exempt because they are not perceived to be goods vehicles).

It would appear that the relevant weight will be the "design weight" but it is not clear whether that is the design weight of the chassis/cab or of the vehicle with the machinery.

Q200. **How would the duty be calculated for my high pressure water-jetting vehicle?**

A. There is, at present, no clear answer to this. It would seem that the tare weight of complete vehicle and equipment is obtained, and duty is then payable on that weight.

Q201. **I have been told that my vehicle is not a goods vehicle, but the authorities refuse to tax it unless I tender it as a goods vehicle. What should I do?**

A. Take advice from a competent source and, if you are satisfied that it is not a goods vehicle, tender the due money for the appropriate category. If the payment is refused, put a note to that effect in the disc holder. The issue of a tax disc is simply a receipt for money, it is not evidence as to the category of vehicle.

Q202. **I want to down rate my vehicle. Can I reclaim the difference in duty?**

A. Yes. You apply for a new tax disc using form VTG10 and any overpayment of duty that results from the change to the vehicle will be refunded to you.

Q203. **My vehicle has been off the road untaxed for some time. Will I have to pay back duty to the point at which it was last taxed?**

A. An enormous amount of duty was lost as a result of people running untaxed vehicles and, when they were caught, paying duty only for the month of the offence. It was, of course, very difficult for the prosecution to show earlier use.

Sections 29–31 of the **Vehicles Excise and Registration Act 1994** create a continuous liability for duty unless notice is given of the acquisition of the untaxed vehicle. In a case where notice is given, it appears that duty will be payable either from the date of the notice or of the expiry of the last licence, whichever is later.

For taxing purposes, time will not be deemed to run while the vehicle is in possession of a motor dealer and is either in a showroom or being used under trade plates.

Q204. **My vehicle has a mechanical fault and is temporarily parked at the side of my house. Do I have to tax it?**

A. Yes. This is a cause of a large number of prosecutions. Section 29 of the **Vehicles Excise and Registration Act 1994** imposes a duty to licence a vehicle which is used *or kept on* a public road. The fact that it is broken-down will rarely avoid the liability for tax, nor will the fact that it is not "on the road" if there is direct access from the road to the place at which it is parked.

Q205. I have paid for the new tax but the disc has not arrived. Can I alter my old one?

A. No. Any alteration will be either forgery or fraudulent use. The penalty for this is a fine and/or up to two years' imprisonment.

Q206. I have an articulated tractor plated to 38,000kg which will pull engineering plant. Do I need to tax it at 38,000kg?

A. No. In a recent case a court determined that the water, sand and cement carried on a "spot mix" vehicle were not "goods or burden". While the position is still a little unclear, in this case the articulated tractor should be taxed as a "haulage vehicle" as described in schedule 1 (paragraph 7) of the **Vehicles Excise and Registration Act 1994**.

It is important to remember that if at any time the vehicle is used to carry goods (ie used with an ordinary articulated trailer) it must be taxed at the appropriate goods vehicle rate.

Q207. I used a special types vehicle to bring a small piece of equipment back for my own use. Is this unlawful?

A. There is no offence if the piece of equipment is for your personal use rather than for use in connection with your trade or business, and is not being carried for hire and reward.

In the event that it is a trade use, then the test may be one of "normality". If it can be shown that this use was a "one-off", you would not expect a prosecution to follow.

Q208. I have a spare vehicle. If I get a trade licence can I use this vehicle if one of my taxed vehicles is broken-down?

A. No. Trade licences are covered by ss.11, 12 and 13 of the **Vehicles Excise and Registration Act 1994**, which authorises the issue of trade licences in very specific circumstances. The law was substantially changed to try to control abuses of the use of these licences.

There are two categories of user:

- motor traders who can use "plates" for vehicles temporarily in their possession or, if they manufacture vehicles, for testing and research, and

- vehicle testers who can use "plates" to carry loads for test purposes (the load being returned to the point of loading) or to take vehicles to and from a test.

Normally a company that maintains, repairs and tests its own vehicles will be entitled to apply for "plates" as a vehicle tester.

Note, however, that "plates" obtained as a motor trader do not allow a vehicle to be used unless it is only temporarily in your possession.

"Motor trader plates" are generally inappropriate for a commercial vehicle operator. Certainly, trade plates cannot be used on a vehicle carrying goods in the ordinary course of the operator's business.

Q209. **I have been offered a disc that still has time to run at a price that will save me money. Should I buy it?**

A.   Absolutely not. Even if the disc is neither stolen nor forged you will be committing a serious offence.

Q210. **My disc was accidentally destroyed. I have applied for a new disc — can I run the vehicle in the meantime?**

A.   Yes. You can either put a note in the disc holder to explain the position or leave the holder empty. Do *not* use a copy disc or a disc taken from another vehicle in any circumstances.

Q211. **I have applied for a tax disc and paid for it, but it has not arrived. Can I use the vehicle?**

A.   Yes. You can either leave the disc holder empty or put a note in the holder instead. In this case though you *must* be able to show that the money has been paid. Proof of posting the cheque, which would require both the proof that the cheque had been issued and that it had been posted, would be the minimum requirement. Ideally, you should be able to show that the cheque had been cleared through your bank.

Q212. **We are thinking of offering a combined road/rail service. Are there any tax benefits?**

A.   Yes. Section 20 of the **Vehicles Excise and Registration Act 1994** anticipates that regulations will be made providing for a rebate provided that certain conditions are met.

INSURANCE

Q213. Where can the law relating to motor insurance be found?

A. In Part IV of the **Road Transport Act 1988**.

Q214. In applying for insurance cover what has to be disclosed?

A. All insurance applications are what is known in law as *Uberrimae Fidei*, which means "of the utmost faith". Thus every question on the proposal form must be answered truthfully and accurately. Any non-disclosure could invalidate the entire policy.

Q215. If drivers have convictions on their ordinary driving licence should these be disclosed?

A. Yes. Failure to disclose convictions is non-disclosure and, in particular, drivers who have a drink-driving related conviction can expect an increased premium.

Q216. If a particular class of insurance is refused by an insurance company can I apply to another company?

A. Yes, but you must disclose that you have been refused cover by another company.

Q217. Is a driver's driving entitlement relevant to insurance?

A. Yes. The operator must be satisfied that the driver has the correct vocational entitlement to drive the goods vehicle that he or she is employed to drive. Every insurance policy is only valid provided that the driver holds a current driving licence and in the case of LGVs that includes the vocational entitlement.

Q218. Should I read all the small print on my policy?

A. Yes, because there may be some limitations, eg age of driver, which would render the policy invalid in certain circumstances.

Q219. What is a "traders' policy"?

A. This is a cover-all policy for all vehicles in a fleet and can include private motor cars. Most insurance companies will require detailed specifications of the vehicles to be covered and for an extra premium will cover vehicles not owned by the operator, but used by him or her, eg hired vehicles. However, this cover will only be third party unless special arrangements are made.

Q220. **If a premium is not paid what happens to the policy?**

A.   It becomes void and the operator will be using the vehicle without insurance. This is an endorsable offence — ie penalty points can be ordered to be placed on the operator's own driving licence and the driver can also be so penalised. Endorsable offences also carry disqualification and a fine if the court so orders.

Q221. **Can a limited company be penalised in the same way?**

A.   No, because a limited company cannot hold a driving licence. However, it can still be fined.

Q222. **Is it advisable to seek the help of an insurance broker in arranging insurance?**

A.   Not necessarily, but most large operators do become clients of insurance brokers who earn their fees by way of commission from the insurance companies. Using the services of an insurance broker is a sure way of getting the exact cover that you require and the vast experience of insurance brokers is such that they will draw your attention to cover that you may require, but had not even thought of.

Q223. **As all motor vehicles have to be insured by law what is the minimum requirement that the law demands?**

A.   The minimum requirement, by law, is for the user of any motor vehicle used on a public road to be insured against third party risks.

Q224. **What are "third party risks"?**

A.   The third party is the "other" person involved in an incident and the cover extends to compensation for bodily injury to that person or persons, damage to their vehicle or property and the cost of any medical treatment. Passengers being carried in a vehicle are also included in the term as a "third party".

Third party risks can be extended to include "fire and theft" cases.

Q225. **How do I obtain insurance cover for repairs following an accident to my own vehicle?**

A.   If you wish to insure for the accidental repairs to your own vehicle, then you must obtain a "comprehensive" policy, which gives you the benefits of third party cover as well as allowing you to claim for repairs to your own vehicle.

Q226.   How do I decide between third party and comprehensive cover?

A.   Insurance quotations are based on a number of elements, not least being claims experience and claims frequency.

All insurers look at the frequency of accidents and the cost of individual accidents, together with the total cost. Depending upon the insurer, the claims experience over a period of three years is examined to obtain an overall view of the "risk" and the premium is then fixed in accordance with the insurer's assessment. It follows, therefore, that the more accidents, the higher the premium.

With the larger fleet, it is usual to have third party cover only, as the cost of arranging comprehensive cover becomes prohibitive. The insurer will almost certainly increase the premiums the following year as a result of claims submitted for accidental repairs to the fleet in the current year. However, most hire purchase and lease hire companies insist on fully comprehensive insurance and evidence of such a policy *must* be submitted to them.

Comprehensive cover may be the most appropriate for smaller operators, as the cost of accidental repairs and time spent waiting for the repairs to be done are factors which the small operator might find unacceptable.

Q227.   How could I reduce my premiums?

A.   Reductions in premiums may be achieved with comprehensive cover which incorporates an excess level. In essence, this means that the excess level is deducted from the repair cost, eg for a £1000 repair with an excess level of £100, the insurers will pay only £900 of the repair cost. The "uninsured" loss, in this case £100, will have to be paid by you, or if the accident was the fault of the other party the excess will form part of the claim for uninsured losses.

Q228.   How can I recover the uninsured losses?

A.   If the accident is not your fault, the uninsured loss can be reclaimed from the other party involved, together with any other "uninsured" losses that you may have sustained, eg hire charges incurred whilst your vehicle is off the road for repairs, injuries sustained by the driver (which will form a separate claim), loss of use and inconvenience and perhaps even loss of revenue until a replacement vehicle is obtained.

Write to the third party to obtain details of his or her insurers if you do not have these already. You will then need to write directly to the third party's insurers, outlining your claim clearly and in full. This service is often provided by brokers at no extra cost and it is often advisable to seek the services of a solicitor whose costs if successful may also be recovered. Some costs may be payable by the claimant as not all costs incurred in any matter are necessarily fully reclaimable. This point should be discussed with the solicitor when he or she is first instructed so that there can be no misunderstanding.

Q229. What do I do if the other party ignores my letters?

A.   If you are satisfied that you have the correct details of the other party and that he or she is deliberately ignoring your letters, then you can ask your broker, insurers or solicitor for help to trace the other party's insurers. If this fails you can, of course, ask for help at your local police station.

Alternatively, you can issue a summons at the County Court.

Q230. How do I issue a summons at the County Court?

A.   In practice, claims for damages above £1000 are generally dealt with through a solicitor, while those below £1000 can be effectively dealt with by the individual, saving time and costs.

Under the County Court rules, claims below £1000 are the subject of proceedings in the small claims court and, in most cases, costs are not allowable.

Q231. What is the procedure in a small claims court?

A.   The parties involved appear before a district judge and the hearing is quite informal. The district judge will listen to the arguments of the two parties (the plaintiff and defendant respectively) and any witnesses who are called, before making his or her decision. The award is made to whichever party is considered to be most deserving, based upon the evidence heard.

The order, when made, constitutes a judgment of the County Court and is enforceable as such.

Q232. What is "goods in transit insurance"?

A.   Goods in transit insurance is not a legal requirement and, as its name suggests, indemnifies the carrier, within the terms of his or her insurance, against claims by the sender or receiver of the goods.

It is normal for carriers to impose conditions upon which they agree to collect and deliver the goods of their customer and which restrict the value of any subsequent claim to these conditions. In the main, this restriction is based upon the weight of the item although, subject to agreement, it can be based on its replacement value.

Q233. How are customers made aware of a carrier's conditions?

A. To ensure that a claim can only fall within the conditions under which the carrier carries goods for the customer, it follows that the customer must be well aware of those conditions by the time a contract is drawn to carry those goods. Thus it is quite common that those conditions appear on rate cards, invoices, etc.

Q234. Is wider cover available?

A. Wider cover is usually available to the replacement value but, again, this is normally agreed between the customer and the carrier. If carriers intend to seek indemnity from an insurance company, they must ensure that the company is notified and agrees to offer indemnity in the event a claim is made.

Q235. Can an insurance company impose additional conditions?

A. As with motor insurance, a carrier faced with an adverse claims experience or a particularly valuable article, may well find that conditions in relation to overnight storage and packaging, for example, will be imposed.

Q236. What advice can be given to a driver in respect of vehicle and load security?

A. Theft of and from motor vehicles has now become endemic. Though it is the car driver who suffers most in terms of numbers, the haulier is by no means exempt.

The driver should take the following basic precautions.

- Where a vehicle has to be left unattended, it must be immobilised by removing the ignition keys and securely locking all windows, cab and body doors and shutters. This should apply if the vehicle is left only for a few minutes. Insurance cover may be invalid if these elementary precautions are not taken.
- Easily removable items, such as a driver's personal radio, television, jacket, etc should, wherever possible, be hidden from view.
- Vehicles parked overnight and away from a depot should, if possible and practical, be left in a security-controlled area. If this is not possible, then choose a well-lit area in a prominent position.
- If the vehicle is being parked during the day, it should be left in a prominent position and, if possible, in view of the driver or members of the public.

- Always ensure that rear or side doors are securely locked when moving and when the vehicle is left unattended — even for only a few minutes.
- Never give lifts to strangers — many companies strictly prohibit the carrying of unauthorised passengers.

In the event that you are unfortunate enough to be the victim of a theft, then report the incident immediately to the nearest police station and contact your depot for advice.

Q237.   Do I need passenger liability insurance?

A.   Passenger liability insurance is compulsory. The cover will extend to authorised passengers, other non-fare paying passengers, "unauthorised" passengers such as hitchhikers and other people who are given lifts. However, it does not cover employees of the owner and operator of the vehicle, as they are covered under an employers' liability policy. Displaying a sign in a vehicle which states "no passengers" or "no liability" does not fully cover the operator or driver against claims by unauthorised passengers for injury or damage resulting from negligence, even if they agree to travel at their own risk negligence.

Q238.   What must a driver do after being involved in an accident?

A.   Drivers must stop in the event of being involved in an accident in which:
- personal injury is caused to any other person
- damage is caused to any other vehicle
- damage is caused to any specified animal other than animals carried on his or her own vehicle
- damage is caused to roadside property.

The driver must give the following information to anybody who has reasonable grounds for requiring the information either at the scene of the accident or later:
- the driver's name and address
- the name and address of the owner of the vehicle
- the vehicle registration number.

The driver should not make any statement at the scene of the accident, particularly admitting or indicating liability. Details of the accident must be reported to the police within 24 hours.

Q239.   What is a "specified animal"?

A.   A specified animal means any horse, ass mule, cattle, pig, goat or dog.

Q240. What is "roadside property"?

A.   Generally, anything which is built on, attached to or growing on the road
and land next to the road. This includes obvious things, such as lampposts
and street signs and less obvious items, such as trees.

Q241. How do I go about making a claim?

A.   The insurance company must be notified immediately of an accident to
any vehicle for which they supply cover. They will also require an accident
report and claim form to be completed, normally within seven days.

The accident report should give as much detail as possible, including:

- time, date and place
- weather and road conditions
- the position of his or her vehicle on the road
- its speed prior to and at the time of impact
- the direction of travel of all the vehicles involved
- the location of identifying objects
- names and addresses of any other parties involved and of any witnesses
- anything that any of the parties may have said which could be helpful to the insurers.

The report form will also normally require a description of events leading
up to the accident, a sketch of the relative positions of the vehicles involved
both before and after the collision and an indication of the damage to
vehicles and property.

HEALTH AND SAFETY

Q242.   What is the "Six Pack"?

A.   This is the colloquial name for a set of six regulations which were implemented on 1 January 1993. They are:

- the **Health and Safety (Display Screen Equipment) Regulations
  1992** (SI 1992 No.2792)

- the **Management of Health and Safety at Work Regulations 1992**
  (SI 1992 No.2051)

- the **Manual Handling Operations Regulations 1992** (SI 1992
  No.2793)

- the **Personal Protective Equipment at Work Regulations 1992** (SI
  1992 No.2966)

- the **Provision and Use of Work Equipment Regulations 1992** (SI
  1992 No.2932)

- the **Workplace (Health, Safety and Welfare) Regulations 1992** (SI
  1992 No.3004).

Q243.   Where can I obtain a copy of the "Six Pack"?

A.   The regulations can be purchased from bookshops stocking HMSO publications and should be available in any main library. Guidance notes to
the regulations can be obtained from HSE Books, PO Box 1999, Sudbury,
Suffolk CO10 6FS (tel: 01787 881165).

Q244.   Is there an overall guide to the regulations?

A.   Approved codes of Practice are available for each of the regulations in
the six pack. They can be obtained, singly or as a six volume pack, from
HSE Books (address as above).

Q245.   How do these regulations affect me as an employer?

A.   All these regulations impose duties on employers in relation to their own
safety, the safety of those whom they employ and the safety of those who
may be affected by the work that is being done.

Although the emphasis may vary, it is likely that all the regulations will
affect all employers. It is therefore important for all employers to have some
knowledge of these regulations.

Q246. I am self-employed. Am I affected by the regulations?

A.   Yes. You have a duty to take care of your own safety and, of course, the safety of those who may be affected by your work. Where five people or fewer are employed, some assessments required by the regulations do not need to be recorded in writing but, as a matter of good practice, it is best to keep a written record.

Q247. What are the "COSHH" regulations? Are they the same as the "Six Pack"?

A.   COSHH stands for the **Control of Substances Hazardous to Health Regulations 1994** (SI 1994 No.3246) which have replaced the 1988 regulations of the same name. Every employer should have implemented those regulations by now. The COSHH regulations are entirely different to the Six Pack regulations, although they introduced the concept of *assessment*. COSHH requires that the risk presented by any substance used in a workplace be assessed by employers and steps taken to remove that risk. If the substance cannot be replaced, all reasonable steps have been taken to reduce the risk and it is still dangerous, then protective equipment should be provided to the employee.

       COSHH only creates a duty to protect those who, not being employees, may be affected by a substance used in the particular work, eg passers-by who might be harmed by dust or contaminated spray.

Q248. I have a small company. Do I need to worry about all these regulations?

A.   You certainly need to know about them. To put it in perspective, the penalties that can be imposed by magistrates are a fine of £20,000 and/or six months' imprisonment; a Crown Court can impose up to two years' imprisonment, with an unlimited financial penalty.

       Additionally, anyone can be prosecuted who is connected to the offence. For example, an injured employee can be prosecuted if the injury was caused by the employee's failure to comply with the regulations, as can the chairperson of a company if there is a direct line of responsibility to him or her.

       When it comes to implementation of the regulations, however, they can be prioritised. The general view is that, in terms of a need to apply them, they should be considered in the order set out in the answer to Question 247 above.

Q249. **I understand that assessments of risk have to be carried out. Who would be best suited to undertake these?**

A.   Assessments have to be carried out by a "competent person". The test of a competent person is pragmatic. The person should have sufficient experience of the particular industry and work for an impartial observer to be satisfied that he or she is competent to make the particular assessment. Formal qualifications may be of assistance, but on their own they may not be enough to make a person competent — practical experience is always helpful.

Choosing the correct person is important because an incompetent assessment may amount to an offence under the regulations. In the context of the manual handling regulations some medical support may be helpful.

Q250. **Will my insurance company help regarding assessments of risk and, if I am insured, do I need to bother with assessments in any case?**

A.   You should certainly discuss the position with your insurers, if only to be satisfied that your policies will cover you in the event of a claim being made. It seems unlikely that insurance companies will be proactive in the sense of providing direct advice, but they will advise as to whether you have done enough for them to be content to continue the cover.

It would seem that employers' liability insurance is not going to be easy to obtain and, certainly, those employers who have not reacted to the new regulations may be uninsurable.

Q251. **One of my staff is pregnant. Is she or her child in any danger if she continues to use a VDU?**

A.   According to research undertaken, there is no particular risk, although there remains a commonly-held belief that pregnant VDU users may be at risk. However, the display screen equipment regulations must be followed.

Q252. **I am about to alter the building in which we work. Are there any problems I should be aware of?**

A.   Yes. One of the most important sets of regulations is the **Workplace (Health, Safety and Welfare) Regulations 1992** (SI 1992 No.3004). These regulations list more than 20 separate areas which are covered by the new requirements.

The regulations came into effect on 1 January 1993 although there are transitional provisions. Workplaces used for the first time after 31 December 1992 were immediately affected, otherwise you have until 1 January

1996 to get your workplace in order. However, any modification, extension or conversion coming into use after 31 December 1992 must comply with the regulations.

Q253. How is "workplace" defined?

A. The interpretation of the word "workplace" is very wide. It includes any place within the premises to which any person has access while that person is at work. It also includes any room, lobby, staircase, road or other place used as a means of access to or egress from the premises, other than a public road.

Q254. Will I need professional help to ensure that I am complying with the workplace regulations?

A. Some of the sections set out the requirements of the regulations in clear terms. In the case of room dimensions, for instance, clear measurements are given. In other sections, objective assessments are required to be made. For example, regulation 12 dealing with traffic routes may well require the assistance of a chartered engineer, surveyor or architect.

Q255. A number of the people I employ have been with me a long time and are totally familiar with the tools they are using. Do I need to train them?

A. Yes. The duty to train is a positive duty and runs through all the recent regulations starting, of course, with COSHH. Long experience often produces bad practice and it is a worthwhile exercise for all to review procedures in the use of tools in the workshop.

Training should be recorded on personnel files and reviewed at appropriate intervals.

Q256. What are "tools" for the purposes of the regulations?

A. The **Provision and Use of Work Equipment Regulations 1992** (SI 1992 No.2932) defines "work equipment" as any machinery, apparatus or tool and any assembly of components which, in order to achieve a common end, are arranged and controlled so that they function as a whole. This means that anything from a hand drill to a bottling plant is encompassed within the regulations. The guide to the regulations sets out a list of examples of work equipment, which ranges from a drill bit to a combine harvester.

Note: the term includes motor vehicles which are not privately owned.

Q257. **The machine that I have just bought is provided with guards. Do I have any additional duty?**

A.   Yes. You must satisfy yourself that the machine is safe for use by your operatives. Guards may have been provided against standards that are not as high as those required by the EC. Additionally, the particular use to which you put the machine may create different problems. Your insurers may be prepared to send an engineer to assess the adequacy of the guards.

Q258. **We make the workers responsible for ensuring that the tools they use are in good condition. Should we confirm that instruction in writing?**

A.   As part of the training that you should provide it is fair that you should require employees to accept that responsibility. However, that will not remove your responsibility to check the condition of any equipment used in your workplace. This requires periodic inspections and, of course, appropriate maintenance to be carried out by someone who is properly qualified.

Q259. **I have done my manual handling assessments, but cannot control my drivers while they are away from the premises. Would I be responsible if a driver was injured?**

A.   Possibly. Clearly, if you have done your assessments you should also have trained the drivers and have a record of the training. Arguably, although it is too soon to be certain, you should advise your customers of the safe weights that your drivers can handle. This would certainly be a duty where a regular customer is involved. In short, if you realise that the driver may be faced with a "need to handle weight", you should check the position with the customer.

Q260. **I have a female driver. Should I ask her if she is pregnant as she will have to lift some weight?**

A.   No. This is an offensive question and is potentially discriminatory. What you should do is to advise her of the risk that might arise if she *were to become* pregnant and invite her to discuss the position with you *should that eventuality arise.*

Q261. **What are considered to be the driver's responsibilities with regard to ensuring the safety of the vehicle?**

A.   When a driver enters and drives a company-owned vehicle, that driver is responsible for the vehicle and its load and must comply with all aspects of current legislation.

Before the vehicle is driven, the driver must, each day, check the following:

- fuel
- oil
- radiator
- screen washer bottle
- operation of lights including indicator, brake and reversing lights
- tyre wear, condition and pressures
- that the number plate is clean and, in the case of a trailer, that the correct plate is fitted
- adjustment of mirrors
- the correct operation of ancillary equipment, eg a tail lift
- that any articulated or draw bar trailer is correctly coupled
- that the vehicle load is safely stowed and secured and that the driver is satisfied that the vehicle is not overloaded, either by gross or axle
- that the interior of the cab is clean and free of litter, with no loose items which may obstruct the use of pedals and controls (loose items often found in cabs, eg radios and portable televisions, should be secured in the event of possible emergency braking)
- that the tachograph is operating correctly and the clock is showing the correct time
- that the lights and reflectors are clean and that all rear markings are clean and in good condition
- that the exterior is clean and free from damage which may hinder the safe use of the vehicle.

Any defects found in the above list (which is not intended to be exhaustive) should be reported immediately for the appropriate action to be taken.

Q262. If an employee who has been with me for less than two years leaves, or is dismissed, claiming it was for health and safety reasons, can he or she take me to an industrial tribunal?

A. Yes. Where an employee can persuade a tribunal that he or she left or was dismissed for health and safety reasons, the tribunals have a specific discretion to deal with cases within the two year period.

CARRIAGE OF DANGEROUS GOODS

Q263. Where can the law, in relation to the carriage of dangerous goods, be found?

A.   In the following:

- the **Road Traffic (Carriage of Dangerous Substances in Road Tankers and Tank Containers) Regulations 1992** (SI 1992 No.743)
- the **Road Traffic (Carriage of Explosives) Regulations 1989** (SI 1989 No.615)
- the **Road Traffic (Carriage of Dangerous Substances in Packages etc) Regulations 1992** (SI 1992 No.742)
- the **Road Traffic (Training of Drivers of Vehicles Carrying Dangerous Goods) Regulations 1992** (SI 1992 No.744)
- the **Chemicals (Hazard Information and Packaging) Regulations 1993** (SI 1993 No.1746)
- the **Carriage of Dangerous Goods by Road and Rail (Classification, Packaging and Labelling) Regulations 1994** (SI 1994 No.669).

Q264. What are the main responsibilities of the operator when dangerous goods are being carried?

A.   Operators must ensure that:

- vehicles are suitable for the load to be carried
- information has been obtained about the nature of the loads from the consignors
- drivers have been given information about the loads *in writing*
- drivers have been provided with suitable training
- precautions are taken during loading, stowage and unloading and to prevent fire and explosion
- vehicles have orange marker plates.

Q265. What are the main responsibilities of drivers when dangerous goods are being carried?

A.   Drivers must:

- keep information about their loads in their cabs including a *Tremcard* (a "transport emergency card")
- display orange marker plates or orange hazard warning panels (tankers and tank containers) when required

- park their vehicles safely
- give information to the police, Department of Transport traffic examiners, fire brigade officers and health and safety inspectors, when required to do so
- take precautions against fire and explosion during loading, stowage and unloading.

Q266. What substances are covered by the regulations?

A. The regulations cover substances which are defined as being:

- dangerous for supply (ie those presenting a danger when being used) — these are the responsibility of the supplier of the product, and
- dangerous for conveyance (ie those presenting a hazard during carriage by road) — these are the responsibility of the consignor of the product.

Q267. Is there a laid down definition of the term "dangerous substance"?

A. Yes. A dangerous substance is defined as a substance in the "Approved List", plus any substance which, by reason of its characteristic properties, creates a risk comparable with the risk created by substances in the Approved List. Those properties relate to substances which are *very toxic, oxidizing* or which are classed as *irritants.*

There are three different grades of "dangerous":

- Packing Group I — very dangerous
- Packing Group II — medium danger
- Packing Group III — low danger.

A substance is "dangerous for conveyance" if it is defined in the Approved List or has characteristics as set out in schedule 2, part 1 of the regulations. Where a dangerous substance is not included in the Approved List it must not be carried. It may not be on the Approved List because it is too dangerous to travel by road or because it is a new substance that has not yet been classified.

Q268. What is the "Approved List"?

A. There are two Approved Lists. The first is the *Approved Substance Identification Numbers Emergency Action Codes and Classifications for Dangerous Substances Conveyed in Road Tankers and Tank Containers* and the second is the *Information Approved for the Classification, Packaging and Labelling of Dangerous Substances.*

Q269.   Where can I obtain copies of the Approved Lists?

A.   They are produced by the Health and Safety Executive (HSE) and can be obtained, along with the operational Codes of Practice, the Code of Practice on Road Tanker Testing and guides to the Regulations, from HSE Books, PO Box 1999, Sudbury, Suffolk CO10 6FS (tel: 01787 881165).

Q270.   Is it an offence not to comply with the Approved Code of Practice (ACOP)?

A.   Not as such. However, the ACOP places legal obligations on operators, drivers and consignors. Any failure to comply with the ACOP could be used during criminal proceedings as evidence of a contravention of the regulations. It would then be up to the person charged to satisfy the court that the regulations had been complied with in some other way, and that could be very difficult.

Q271.   What is the definition of "explosives"?

A.   Explosives are defined as explosive articles or substances which have been classified under the **Classification and Labelling of Explosives Regulations 1983** (SI 1983 No.1140) as being either in Class 1 or those which are unclassified. An *explosive article* is an article that contains one or more explosive substances. An *explosive substance* is a solid or liquid substance, or a mixture of solids and liquids, capable by chemical reaction of producing gas at a temperature, pressure and speed as to cause damage to surroundings, or which is designed to produce an effect by heat, light, sound, gas or smoke, or a combination of them, as a result of a non-detonative, selfsustaining exothermic chemical reaction.

Q272.   What is meant by "conveyance by road"?

A.   Conveyance by road means the period from the moment loading of the vehicle begins until the moment the vehicle has been completely unloaded and all substances have been removed, including contaminated drums and other packages, or, in the case of tankers, the tank has been cleaned or purged.

Q273.   What is meant by "carriage" with regard to explosives?

A.   Where explosives are being carried, carriage means from the commencement of loading explosives into the vehicle or trailer until the explosives have been unloaded, whether or not the vehicle is on the road at the time. Where the explosives are loaded on an unattached trailer or semi-trailer, carriage begins and ends when the trailer is attached to, and later detached from, the towing vehicle, or when the explosives have been unloaded, whichever is the sooner.

Q274. Do substances always require labelling?

A. Substances dangerous for conveyance are nearly always dangerous for supply, but the reverse is not always true. Therefore, substances may require labelling for supply purposes, but not for conveyance.

Q275. What are the main features of the supply label in general?

A. These are:

- the name and address of the manufacturer, importer, wholesaler or other supplier
- the designation of the substance, usually the chemical name
- the indication of risk, eg may cause fire, and the appropriate symbol
- the safety phrase, eg "keep container dry".

Q276. What are the main features of the label for conveyance by road?

A. These are:

- the name and address of the consignor or telephone number (or both) or details of another person in the UK from whom expert advice can be sought in the event of an emergency
- the designation of the substance, usually the chemical name
- the substance identification number
- the appropriate warning sign from the Approved List
- where the quantity in any receptacle exceeds 25 litres, details of the dangers created by the substance and the action to be taken in the event of an emergency involving the substance.

Q277. Who is responsible for complying with the law when dangerous goods or explosives are being carried?

A. The operator and the driver.

Q278. Who is regarded as the operator?

A. The operator is defined as the holder of the Operator's Licence or the registered keeper in respect of vehicles not subject to operator licensing (where the vehicle is subject to a hire purchase or lease hire agreement that company may be the registered keeper; in that event it is the user who is the operator). The operator of a tank container is defined as the owner or his or her agent or, in their absence, the operator of the carrying vehicle.

Q279. Are the terms "road tanker" and "tank container" defined?

A. Yes. A *road tanker* is defined as a goods vehicle with a tank structurally attached to, or which is an integral part of, the frame of the vehicle. A *tank container* is defined as a container having a capacity greater than three cubic metres.

Q280. Do road tankers, tank containers and vehicles used for the carriage of dangerous goods and explosives have to comply with any specific requirements?

A. Yes. Road tankers and tank containers must be:
- properly designed
- of adequate strength
- of good construction
- constructed from sound and suitable material which would not be adversely affected by contact with the load
- designed, constructed and maintained so that the contents cannot escape
- suitable for the purpose
- equipped with adequate and correct ancillary fittings.

Vehicles used for carrying packaged dangerous goods must be properly designed, constructed and maintained. Specialised vehicles are not required but they must be suitable for the purpose.

Any vehicle or freight container used for the carriage of explosives must be suitable for the safety and security of the explosives being carried, bearing in mind the type and quantity of the explosives. The maximum quantity of explosives permitted to be carried must not be exceeded. A greater quantity of explosives than the vehicle or container is suitable for must not be carried.

Q281. What are the maximum quantities of explosives that can be carried?

A. The maximum quantities are as follows:
- Division 1.1, Compatibility Group A — 500kg
- Division 1.1, Compatibility Groups B, F, G or I — five tonnes
- Division 1.1, Compatibility Groups C, D, E or J — 16 tonnes
- Division 1.2, any Compatibility Group — 16 tonnes
- Division 1.3, any Compatibility Group — 16 tonnes
- unclassified explosives being carried solely in connection with an application for their classification — 500kg.

Different compatibility groups of explosives must not be carried together unless permitted by the regulations.

Q282. What are the testing and certification requirements for road tankers, tank containers and vehicles used to carry dangerous goods and explosives?

A. The *Road Tanker Testing Code of Practice* requires periodic testing and certification. The certificate must indicate the substances that may be carried. Tankers must be retested and certified following accident damage, repairs or modification.

Q283. When do the Road Traffic (Carriage of Dangerous Substances in Road Tankers and Tank Containers) Regulations 1992 (SI 1992 No.743) not apply?

A. The regulations do not apply to vehicles carrying non-dangerous bulk powder, granules or liquids. They also do not apply to the carriage of dangerous goods when vehicles are:

- engaged on international transport under the ADR regulations (Agreement Concerning the International Carriage of Dangerous Goods by Road) or under the CIM regulations (the International Convention for the Conveyance of Goods by Rail)
- under the control of the armed forces or visiting forces
- exempt from paying vehicle excise duty under the **Vehicles Excise and Registration Act 1994**
- road construction vehicles other than road tankers, used for conveying liquid tar.

Q284. What information should be available before dangerous goods or explosives are carried?

A. Before dangerous goods or explosives can be carried, operators must obtain from consignors sufficient information to ensure that the law is complied with and that the hazards of the substance being carried are known. Consignors have a duty to supply such information.

The information should include:
- the identity of the substance being carried
- the nature of the hazards
- the action to be taken in an emergency
- what protective devices should be used to deal with any spillage
- advice on first aid

- a contact telephone number, and
- in the case of dangerous goods carried in packages, the quantity and size of the receptacles carrying the dangerous goods.

Drivers must be given written information, which must be kept inside the cab and be secured in such a way so as to be readily available at all times. This information must give details of the substances on board, their associated hazards and any emergency action that has to be taken in the event of an accident. The written information can be in any form, but is often in the form of a *Tremcard* ("transport emergency card").

When carrying classified explosives, operators must give drivers written information concerning:

- the Division and Compatibility Group
- the net mass, or gross mass, of each type of explosives carried
- whether the explosives carried are "articles" or "substances"
- the name and address of the consignor, the vehicle operator and the consignee, and
- any other information necessary to enable drivers to be alerted to the dangers which might arise and the emergency action to be taken.

Q285. How should such information be treated once unloading is completed?

A. Such written information must be removed, destroyed or securely locked away once the dangerous goods or explosives have been unloaded from the vehicle.

Q286. When do the Packaged Goods Regulations (SI 1992 No.742) apply?

A. The regulations apply when:

- toxic gases, organic peroxides, asbestos, asbestos waste and other substances of similar risk are being conveyed in any quantity
- substances in Packing Group I are being conveyed in receptacles that have a capacity of five litres or more
- substances in Packing Groups II and III are being conveyed in receptacles which have a carrying capacity of 200 litres or more
- substances in bulk are being conveyed which are not subject to the **Road Traffic (Carriage of Dangerous Substances in Road Tankers and Tank Containers) Regulations 1992** (SI 1992 No.743)
- self-reactive organic peroxides or flammable solids are being conveyed in any quantity which have to be transported at controlled temperatures

(drivers must have adequate means of ensuring that the temperature is maintained at a safe level)

- LPG (liquefied petroleum gas) is being carried in cylinders with a water capacity of five litres or more, except where it is providing fuel for equipment.

In addition, the regulations apply to empty receptacles that have previously contained specified substances until they have been thoroughly cleaned and purged.

Greater concentrations of substances must not be carried than are shown in the Approved List, or in contravention of conditions relating to inert or stabilising substances.

Q287. **When do the Road Traffic (Carriage of Dangerous Substances in Packages etc) Regulations 1992 (SI 1992 No.742) not apply?**

A. The regulations do not apply to:

- substances used to run the vehicle
- movements under the ADR and CIM Regulations
- the delivery of goods between premises which are considered to be within the "immediate vicinity" of each other
- agricultural vehicles, provided the substances have been suitably diluted
- vehicles that are not goods vehicles
- tankers and tank containers
- radioactive and explosive materials
- foods, cosmetics and medicines as defined in the **Classification, Packaging and Labelling of Dangerous Substances Regulations 1984** (SI 1984 No.1244).

Q288. **What precautions should be taken during loading, stowage and unloading?**

A. Everyone involved in the conveyance of dangerous goods must ensure that their actions do not endanger others and that the law on health and safety is observed at all times. Drivers must ensure that vehicles are not overloaded, that the loads do not overhang the sides of the vehicle and that those substances carried which are likely to inter-react are carefully segregated.

Packages which weigh 500kg or more must be secured individually, unless they have been secured in such a way as to make individual lashing unnecessary. Drivers must ensure that goods are secured in the vehicles in such a way as to minimise the risk of damage to the packages.

Q289.   What precautions should be taken against fire or explosion?

A.   All vehicles carrying dangerous goods should carry a fire extinguisher in the cab. This should contain a minimum of 1kg dry powder, halon or BCF. In some cases it may be necessary to carry an additional extinguisher for special loads.

It is important to remember to check the extinguishers regularly.

Drivers must:

- guard against fire or explosion
- ensure that loading, stowage and unloading is carried out safely
- in general take care, particularly when on the highway and when in the vicinity of the public.

Q290.   When must vehicles carrying dangerous goods in packages be marked?

A.   When the load exceeds 500kg of dangerous substances falling within the regulations, the vehicle must carry and display rectangular orange plates surrounded by a black border, at the front and rear of the vehicle. If the substances are being carried in a trailer and that trailer has been detached with a load still exceeding 500kg, the marker boards must remain on the rear of the trailer.

The marker boards must remain on display until all of the dangerous substances have been removed. They must be kept clean and free from obstruction and must be removed or covered up when no dangerous substances are being carried.

Q291.   What marking is required for vehicles carrying explosives?

A.   Vehicles carrying explosives must be marked front and rear with a reflectorised orange plate surrounded by a black border. A square placard with an orange background and black border depicting a "bomb blast" and any figures or letters denoting classification and Compatibility Group shown in black, set at an angle of 45°, must be displayed on each side of the vehicle container or trailer.

Q292.   What are the rules on overnight parking?

A.   Vehicles carrying dangerous goods must be parked in a safe place or they must be supervised at all times by the driver or a competent person over 18 years of age. This requirement applies to tankers displaying hazard warning signs on which the emergency action code ends with the letter E. It does *not* apply if the driver can show that:

- the tank is empty, or
- if the identification code number 1270 is displayed, no petrol is being conveyed and the tank is empty, or
- if the number 1294 is displayed, no toluene is being carried and the tank is empty.

A "safe place" would be a supervised lorry park or in the open air where the public does not pass or gather, eg a factory yard.

If the driver is unable to comply with any of these requirements, then the vehicle must not be parked on a road, it must not be parked within 15 metres of occupied premises, or where the public pass or gather.

The definition of a "road" for these purposes includes a lay-by, but not a disused road loop or an HGV parking area at a motorway service area.

Q293.   What are the rules for short stops?

A.   Drivers should park their vehicles within their sight or where they can get to them within a few minutes. If drivers are unable to comply with these requirements, vehicles should be supervised by a competent person over 18 years of age.

In the event of a stop being required because of an accident or emergency, drivers should stay with their vehicles.

When drivers leave their vehicles they should always be secured against theft or interference.

Q294.   What steps should be taken when vehicles are carrying explosives?

A.   Operators, drivers and other persons involved in the carriage of explosives must take all reasonable steps to prevent accidents and minimise the harmful effects of any accident. They must prevent unauthorised access to, or removal of, all or part of the load.

Operators and drivers must ensure that a competent person is constantly in attendance with vehicles whenever the drivers are not present, except during stops within a safe and secure place.

A *safe and secure place* for this purpose is:

- a factory or ammunition store licensed under the **Explosives Act 1875,** or
- a place with an exemption certificate under the **Explosives Act 1875 (Exemptions) Regulations 1979** (SI 1979 No.1378), or

- a site where adequate security precautions exist.

Q295. **Are there any restrictions on the routes taken by vehicles carrying explosives, on the duration of the journey and delivery?**

A. Yes. A vehicle carrying more than five tonnes of Division 1.1 explosives must follow a route agreed with the police for each area through which it is to pass.

Operators and drivers must:

- ensure, in every case, that the carriage of explosives is completed within a reasonable time, taking account of the distance involved
- ensure that the explosives are unloaded as soon as practicable on arrival
- ensure that the explosives are delivered to the consignees or their agents or to another person who accepts them into custody for onward transmission, provided they are delivered to a safe and secure place or a designated parking area in an airport, a railway transshipment depot or siding, a harbour or harbour area.

If the explosives cannot be delivered as required they must be returned to the consignors or their agents.

If the explosives are loaded in a trailer, the trailer must not be detached except in a safe place or in an emergency.

Q296. **Are there any minimum age requirements for drivers when vehicles are carrying explosives?**

A. Yes. The regulations specify that drivers and attendants of vehicles carrying explosives, persons made responsible for the security of explosives, or those travelling in a vehicle carrying explosives must be at least 18 years old, unless they are in the presence of, and supervised by, a competent person over 18 years old.

Q297. **What are the training requirements for drivers and attendants of vehicles carrying dangerous goods and explosives?**

A. Employers have a duty to ensure that drivers of vehicles carrying dangerous goods receive training so that they understand:

- the dangers arising from the substances they are carrying
- the procedures to adopt in an emergency
- the use of safety equipment
- the loading and unloading procedures, pre-journey checks, vehicle marking and general legal requirements.

Employers must ensure that drivers hold the relevant ADR certificates covering the vehicle being driven and the substances being carried. In addition, they have a duty to provide, and pay for, any necessary training.

Operators of vehicles carrying explosives also have a duty to ensure that attendants as well as drivers have the necessary training.

TRAINING REQUIREMENTS

Since 1 July 1992, the **Road Traffic (Training of Drivers of Vehicles Carrying Dangerous Goods) Regulations 1992** (SI 1992 No.744) require ADR vocational training certificates to be held by drivers of tankers and tank container vehicles carrying dangerous goods over 3500kg gross weight with capacities over 3000 litres, and vehicles carrying explosives.

From 1 January 1995, the regulations require such certificates to be held by drivers of other vehicles carrying dangerous goods.

Q298. What are the current training requirements under the Road Traffic (Carriage of Dangerous Substances in Packages etc) Regulations 1992 (SI 1992 No.742)?

A. Employers are required to either establish in-house training or arrange training with training establishments approved under the National (Dangerous Substances) Driver Training Scheme for Hazpak training.

Q299. How can drivers obtain the required ADR training certificates?

A. The certificates, which are issued by the DVLA at Swansea, are obtained by attending a government-approved course and passing a written examination set by the City and Guilds of London Institute. Candidates who pass the examination receive their certificates direct from the DVLA. Those who fail can apply to re sit the examination without further training within 16 weeks of being notified of their failure.

The certificates remain valid for five years and are renewable subject to holders attending an approved refresher course and taking a further examination.

Further information on the examination and the required training course can be obtained from the Department of Transport (tel: 0171-276 4963).

Q300. What are the current minimum ADR Directive requirements for tanker driver training?

A. The minimum requirements are listed overleaf.

1. Core module of a day and a half, including exercises on first aid and fire fighting.
2. A half-day module on tanker or packaged goods.
3. A three/four hour course on classification of material carried.

Q301. What are the driver's responsibilities in relation to ADR certificates?

A. Drivers must carry their certificates with them when driving vehicles carrying dangerous goods. They must produce them on request to a police officer or a Department of Transport traffic examiner. It is an offence to drive a vehicle carrying dangerous goods when not the holder of a full, transitional or provisional certificate. It is also an offence to fail to produce the certificate when requested to do so by an enforcement officer.

Q302. What action should drivers take in the event of an incident/accident involving dangerous goods?

A. In the event of an incident prejudicing the safety of the load or of the public (eg a traffic accident or a leakage from the load), the first responsibility should be to arrange, where necessary, that the police and emergency services are alerted and that assistance is given to any person who is in any immediate danger or who has suffered injury.

Action by drivers to tackle a fire or contain a spillage should be undertaken only if it is considered safe to do so, having regard to the circumstances of the emergency, the nature of the load being carried and the equipment available on the vehicle.

The driver should:

- move the vehicle to a place where any leakage would cause less harm or damage, if it is safe to do so in the circumstances
- stop the engine
- operate the emergency warning device, if fitted, but only in such circumstances where there is no likelihood of it igniting any flammable vapour which may have escaped
- wear appropriate protective clothing
- keep people away from the scene
- place a hazard warning device in position to the rear of the vehicle, where such a device is carried
- prevent smoking and keep other vehicles out of any fire risk area
- remove from the cab of the vehicle all written information concerned with the load being carried

- inform a senior manager at the depot, by radio or telephone, of the details of the incident and the action that has been taken.

When the police and fire brigade arrive at the scene, they will take charge and drivers should:

- show them any written information they have concerning the substance in question and the rest of the load at that time (it is important that drivers pass to the emergency services only information concerning the load *at that time*; they should keep separate any information relating to the goods that they have already delivered, or that they have yet to deliver, or that they have yet to collect)
- tell them all the information that may be helpful concerning the load
- tell them what action, if any, they have taken prior to the arrival of the emergency services.

The driver should also note that if the incident occurs on some other person's premises, eg at a customer's premises whilst collecting or delivering, or on the car park of a garage, etc, then the owner/manager of those premises has certain obligations under the **Health and Safety at Work, etc Act 1974**. The driver must, therefore, always inform the owner/manager of the exact nature of the incident as soon as possible. In such circumstances, it may well be the case that the owner/manager will want to take charge of the situation and act in accordance with the company's health and safety policy.

The driver should be as helpful as possible in such circumstances and supply the owner/manager of the premises with any information requested concerning the nature of the load.

ENVIRONMENTAL ISSUES

Q303. Where can the environmental provisions relating to goods vehicle operators' licensing be found?

A. In the **Transport Act 1968**, as amended by schedule 4 of the **Transport Act 1982**, s.69a–g and regulations 21 and 22 of the **Goods Vehicles (Operators' Licences, Qualifications and Fees) Regulations 1984** (SI 1992 No.176).

Q304. What is the aim of the environmental provisions in the operators' licensing legislation?

A. The aim is to balance the commercial needs of goods vehicle operators against the adverse effects of their operations on the environment which are capable of prejudicially affecting the use or enjoyment of land in the vicinity of an operating centre.

Q305. What is an "operating centre"?

A. An operating centre is the place where the vehicles authorised on the licence are normally kept when not in use.

Q306. What does a licensing authority examine when considering a licence application under the environmental provisions?

A. The following points will be considered.

- The nature and use of other land and property in the vicinity and the effect the granting of the application will have on the local environment.

- In the case of an existing operating centre, consideration will also be given to whether a grant will result in a material change which would have an adverse impact on the environment, eg an increase in the number of vehicles and trailers based at the operating centre. If there will be no material change, a licensing authority can only refuse an application if he or she is not satisfied about the parking arrangements. Where a licensing authority is precluded from refusing an application because there will be no material change, he or she can only impose conditions on the licence after giving operators a chance to make representations about the effect of the proposed conditions on their business and after giving those representations special consideration.

The licensing authority will also take into account the following.

- Planning permission (in cases where the site has not previously been used as an operating centre).
- The number, type and size of the vehicles to be authorised and the parking arrangements.
- The nature and time of use of the operating centre by the authorised vehicles and trailers, including the use of special equipment relating to the operation of the vehicles.
- How, and how often, authorised vehicles would enter and leave the operating centre.

Q307. **In granting an application, what environmental conditions can a licensing authority impose on a licence?**

A.   There are four types of condition that can be imposed under the regulations. Those that:

- limit the number, type and size of the vehicles and trailers authorised on the licence
- regulate the parking arrangements (including those for the purposes of loading and unloading) in the operating centre and on the surrounding roads
- restrict the hours of operation and maintenance of authorised vehicles and trailers
- specify the direction in which authorised vehicles enter and leave the operating centre.

Q308. **Is there any right of appeal against a decision of a licensing authority to impose conditions or refuse an application on environmental grounds?**

A.   Yes. Operators who are aggrieved by any decision made by a licensing authority can appeal to the Transport Tribunal.

Q309. **What happens if the conditions imposed are breached?**

A.   Operators who breach licence conditions may be prosecuted and, if convicted, fined up to £500. The licensing authority can also take disciplinary action under s.69(1)(e) of the **Transport Act 1968** (as amended) against a licence. In serious cases of breach of conditions, revocation of the licence will almost certainly be considered.

Q310. Do conditions imposed apply to every vehicle using the operating centre?

A. No. They only apply to the vehicles authorised on the licence. They do not apply to:

- vehicles under the operator's licensing limit, or
- visiting vehicles that are authorised at other operating centres belonging either to the licence holder or to other operators.

Q311. Who can object to a licence application?

A. A chief officer of police; a local authority (but not a parish council — this may change in the near future); a planning authority; the British Association of Removers; the Freight Transport Association; the Road Haulage Association; the General and Municipal Workers Union; the National Union of Railwaymen; the Transport and General Workers Union; the United Road Transport Union; and the Union of Shop, Distributive and Allied Workers. Like the licence applicant, objectors have a right of appeal to the Transport Tribunal against a decision of a licensing authority.

Q312. Can local residents object to a licence application?

A. Owners and occupiers of land, including buildings, in the vicinity of an operating centre, who may be adversely affected, have the right to make representations on environmental grounds to the licensing authority.

Representers *cannot* initiate an appeal to the Transport Tribunal on environmental grounds, but if there is an appeal they are given the opportunity of becoming parties to it.

Q313. What procedure should objectors and representers follow?

A. Objections and representations must be made in writing, stating the grounds on which they are made. Objections must reach the licensing authority within 21 days of the publication of the details of the application in the official Traffic Area publication *Applications and Decisions*. Representations must reach the licensing authority within 21 days of the date of the advertisement of the details of the application in newspapers or in a local newspaper (ie one that circulates in the locality). At the same time that objections and representations are sent to the licensing authority, a copy of the objection or representation must be sent to the licence applicant.

Q314.   Does the newspaper have to be a local one?

A.   The Act is confusing in that it refers to "newspapers or a local newspaper". Thus it can be argued that national newspapers circulate throughout the UK; so as the word newspaper is in the plural in the Act it is submitted that an advertisement in two national newpapers would be lawful. This has yet to be argued before the Transport Tribunal. To be safe, one local newspaper (which can be a free paper) which circulates in the area should carry the advertisement.

Q315.   What information has to be given regarding the environmental aspect when making a licence application?

A.   If, after the application has been advertised in newspapers or in a local newspaper, representations on environmental grounds are received from owners and occupiers of land in the vicinity, an applicant will be asked to complete a GV79E form, headed *Supplementary Environmental Information*, which asks for specific information (see box below). It is important that the form is filled in carefully and accurately as a licensing authority will consider it a binding statement of the applicant's intentions.

FORM GV79E — SUPPLEMENTARY ENVIRONMENTAL INFORMATION

The form requires:

- details of the vehicles and trailers to be normally kept at the operating centre

- a sketch map or plan showing where the vehicles are to be parked within the operating centre

- details of other places where vehicles and trailers may be parked in the vicinity of the operating centre

- if the premises are not the applicant's own, evidence that the applicant has permission to use them for the parking of vehicles and trailers

- the hours between which lorries will normally arrive at and leave the operating centre

- whether lorries will normally use the operating centre on Saturdays and Sundays and, if so, between what hours

---

- whether maintenance work will be carried out at the operating centre and, if so, between what hours that work will normally be done

- whether any of the maintenance work will normally be done on Saturdays and Sundays and, if so, between what hours

- whether there are any covered buildings at the operating centre in which to carry out maintenance work

- a plan of the operating centre, showing entry and exit points, the main building, surrounding roads with names, and the normal parking area for authorised vehicles and trailers (the scale of the plan should be indicated, if possible)

- in the case of a site not previously used as an operating centre, information about any application for or grant of planning permission, in relation to the proposed use

- any further information about the environmental aspects of the operating centre which the applicant considers relevant to the licence application.

Q316. What does the licensing authority regard as the "vicinity" of an operating centre?

A.   It is up to the licensing authority to decide the geographical limit of the vicinity with regard to the particular circumstances of each individual case. However, in general terms, being in the vicinity consists of being within sight, sound or smell.

Q317. If objections and/or representations are made and are held to be valid will there be a public inquiry?

A.   Not necessarily. The licensing authority will decide firstly whether or not the objections and/or representations are valid and have been submitted in time. Sometimes it is possible to suggest conditions to the applicant which would meet the objections/representations and if the applicant is agreeable to a grant with a condition(s), then it will not be necessary to hold a public inquiry. If this is not possible, or if there are so many valid points against the grant, the licensing authority will hold a public inquiry to which all valid objectors and representers will be invited to attend and give evidence. Whilst parish councils have no right of objection their officers can be called to give relevant evidence.

CONTROLLED WASTE

Q318. What is "controlled waste"?

A. Controlled waste is household, industrial or commercial waste. Waste is defined as including:

- any substance which constitutes scrap material or an effluent or other unwanted or surplus substance arising from the application of any process, and
- any substance or article which requires to be disposed of as being broken, worn out, contaminated or otherwise spoiled.

Q319. Do I have to be registered to carry controlled waste?

A. Yes. The **Control of Pollution (Amendment) Act 1989** requires carriers of controlled waste to be registered with a waste regulation authority (WRA).

Q320. Are there any exceptions to the registration?

A. The registration requirement does not apply to:

- transport between different parts of the same premises
- transport to a place in Great Britain of controlled waste brought from another country and which does not land in Great Britain until it arrives at that place
- a waste collection authority, waste disposal authority or waste regulation authority
- a charity or voluntary organisation
- the producer of the waste, unless it is building or demolition waste, which is defined as waste arising from works of construction or demolition, including preparatory works.

Consequently, operators can transport their own waste without being registered unless it is waste from a building or demolition site.

Q321. What is the duty of care required when carrying controlled waste?

A. The **Environmental Protection Act 1990** places a duty of care on any person who imports, produces, carries, keeps, treats or disposes of controlled waste to take all reasonable measures to:

- prevent the treatment, keeping or disposal of controlled waste in a manner likely to cause pollution to the environment or harm to human health

- prevent the escape of waste from his or her control, or that of any other person
- secure that on the transfer of waste, the transfer is only to an *authorised person* or to a person for *authorised transport purposes*
- secure a written description of waste which will enable anyone to avoid an unauthorised deposit of controlled waste.

An *authorised person* is:
- a waste collection authority
- the holder of a waste management licence or disposal licence
- a registered carrier of controlled waste
- a waste disposal authority in Scotland, or
- a person coming within specified exemptions.

Authorised transport purposes are:
- the transport of controlled waste between different places within the same premises, and
- the transport to a place in Great Britain of controlled waste brought from an outside country which does not land in Great Britain until it arrives at that place.

Q322. **Is there any other documentation required for the carrying of waste, apart from the written description of waste?**

A.   The **Environmental Protection (Duty of Care) Regulations 1991** (SI 1991 No.2839) require that at the same time as the written description of waste is transferred, a transfer note must be completed and signed by the transferor and transferee of the waste.

A transfer note must:
- identify the waste and its quantity
- state whether it is loose or in a container and identify the kind of container
- state the time and date of the transfer
- state the names and addresses of the transferor and transferee
- state whether the transferor is the producer or importer of the waste
- state to which category of person the transferor and transferee belong.

The categories are:
- a waste collection authority
- a waste disposal authority (in Scotland)

- the holder of a waste management licence or a person exempt from holding one
- a registered carrier of controlled waste or a person exempt from registration.

Registered carriers must give the name of the authorities they are registered with and their registration numbers.

If the transport is for an authorised transport purpose the purpose must be specified.

Q323. For how long should records relating to controlled waste be kept?

A. The transferor and transferee must each keep a copy of the written description of waste and the transfer note for *two years* after the transfer. A waste regulation authority can, by written notice, require production and a copy of the documents.

Q324. Where can controlled waste be deposited?

A. Under the **Environmental Protection Act 1990**, the deposit of "directive" waste on land is prohibited unless the occupier of the land is the holder of a waste management licence which authorises the deposit in question and the deposit is in accordance with any conditions specified in the licence. "Directive" waste is defined as materials which the holder discards (or is required to discard), with certain exemptions. The exemptions to the requirement to have a licence are set out under the **Waste Management Licensing Regulations 1994** (SI 1994 No. 1056). Exempt activities must be registered with the local waste regulation authority.

SPECIAL WASTE

Q325. What is "special waste"?

A. Special waste is any controlled waste which is (or contains) a specified substance which makes it dangerous to life or gives it a flash point of 21°C or less; or is a medicinal product available only on prescription.

Q326. What is the documentation required for the carrying of special waste?

A. Under the **Control of Pollution (Special Waste) Regulations 1980** (SI 1980 No.1709), producers of special waste must prepare copies of a prescribed consignment note before the waste is removed from their premises. Carriers of the waste must complete their section of the consignment note before removing the waste from those premises to the place of disposal.

Except for the disposal authority which collects and disposes of the waste in its own area, carriers must give copies of the consignment note which are supplied to them to the disposer. Carriers must retain one copy on which all the sections have been completed.

Carriers must keep a register containing copies of all consignment notes relating to special waste transfers.

Q327. **For how long should records relating to special waste be kept?**

A. Copies of the consignment notes must be kept in the carriers' register for *at least two years* from the date on which the waste was removed.

TRANSFRONTIER SHIPMENT OF WASTE

Q328. **What is "transfrontier shipment"?**

A. When waste is transported from Great Britain to another country it becomes subject to EU Regulation 259/93 on the supervision and control of shipments of waste within, into and out of the Community and the **Transfrontier Shipment of Waste Regulations 1994** (SI 1994 No.1137).

Q329. **What are the rules and the documentation required for the carrying of hazardous waste?**

A. If waste is destined for disposal to a country outside Great Britain, prior consent must be obtained from the competent authoritites of dispatch (the exporting country), destination (the importing country) and transit (any countries which the shipment passes through). These authorities must subsequently be informed once the disposal has taken place. In Great Britain, the waste regulation authorities (WRAs) are the competent authorities for imports and exports of waste, and the Secretary of State is the competent authority for transfer of waste through Great Britain.

Anyone shipping waste into or out of Great Britain must apply to the WRA for a certificate demonstrating that they have a financial guarantee or equivalent insurance, sufficient to cover the cost of the shipment.

WATER POLLUTION

Q330. **Where can I find the legislation relevant to water pollution?**

A. The **Water Resources Act 1991** makes it a punishable offence to:

- cause or knowingly permit any poisonous, noxious or polluting matter to enter controlled waters
- cause or knowingly permit any matter to enter inland waters so as to tend directly or in combination with other matter, to impede the proper flow of the water of the stream in a manner leading, or likely to lead, to a substantial aggravation of pollution
- to cause or knowingly permit matter, other than trade or sewage effluent, to enter controlled waters by being discharged from a drain or sewer in contravention of a relevant prohibition
- to cause or knowingly permit trade effluent or sewage to be discharged either into controlled waters, or from land in England and Wales through a pipe into the sea beyond the limits of controlled waters
- to cause or knowingly permit trade or sewage effluent to be discharged in contravention of a relevant prohibition from a building or fixed plant onto or into any land, or into the waters of any lake or pond which are not inland waters.

It is, however, permissible to discharge effluent into water where it is in accordance with a consent issued by the National Rivers Authority (river purification authority in Scotland).

The **Water Resources Act 1991** is also concerned with the duties and functions of the NRA, including the control of pollution.

Q331. **What is the position regarding possible water pollution with regard to vehicle maintenance and washing off operations?**

A. Discharges to watercourses, other than those from the surface water drainage system and natural run-off, are subject to consent by the NRA.

Any individual or company wishing to make a direct discharge to controlled waters must apply to the NRA for a consent to discharge.

Q332. **What steps can be taken to avoid water pollution?**

A. See the box below.

HOW TO AVOID WATER POLLUTION

- Have interceptors for oil and grease installed into the drainage system.
- Do not put oil, petrol, diesel fuel or chemicals down drains or into gutters.
- Do not throw rubbish into rivers or streams.

- Do not allow any potentially polluting matter to escape into rivers or streams or the drainage system.

- Do not put rubbish into brooks or on the banks, as this can block river channels and culverts and cause flooding.

- Ask for advice if you are not sure how to dispose of a potential pollutant.

- Report any suspected pollution to the NRA.

- Tell the NRA if there is a spillage of a potential pollutant.

- Consult the NRA before undertaking any work on, under or over, a watercourse or on the banks of a river or stream.

Q333. **What are the penalties for polluting a watercourse?**

A.   Where a watercourse is polluted by a discharge without consent, or by a discharge that has consent but is outside its set limits, or when a contaminating substance enters a watercourse, the person or persons responsible may be prosecuted and, on conviction, fined up to £20,000 and/or given a prison sentence of up to three months per offence in the magistrates' court. The Crown Court can impose an unlimited fine and/or a prison sentence of up to two years.

TACHOGRAPHS AND DRIVERS' HOURS

Q334. Where can the law relating to EC drivers' hours and tachograph legislation be found?

A. In the following regulations:

- EC Regulations 3820/85 3821/85 and 3314/90
- the **Community Drivers' Hours and Recording Equipment (Exemptions and Supplementary Provisions) Regulations 1986** (SI 1986 No.1456)
- the **Community Drivers' Hours and Recording Equipment Regulations 1986** (SI 1986 No.1457)
- the **Drivers' Hours (Harmonisation with Community Rules) Regulations 1986** (SI 1986 No.1458)
- the **Community Drivers' Hours and Recording Equipment (Exemptions and Supplementary Provisions) (Amendment) Regulations 1987** (SI 1987 No.805)
- the **Passenger and Goods Vehicles (Recording Equipment) Regulations 1991** (SI 1991 No.381).

Q335. Where can the rules regarding domestic drivers' hours be found?

A. The domestic drivers' hours rules are contained in:

- the **Transport Act 1968** (Part VI), as amended by the **Drivers' Hours (Passenger and Goods Vehicles) (Modifications) Order 1971** (SI 1971 No.818)
- the **Drivers' Hours (Goods Vehicles) (Modifications) Order 1986** (SI 1986 No.1459)
- the **Drivers' Hours (Goods Vehicles) (Exemptions) Regulations 1986** (SI 1986 No.1492)
- the **Drivers' Hours (Harmonisation with Community Rules) Regulations 1986** (SI 1986 No.1458)
- the **Drivers' Hours (Goods Vehicles) (Keeping of Records) Regulations 1987** (SI 1987 No.1421).

Q336. What is the purpose of the legislation?

A. To keep tired drivers off the road — they are safety measures.

Q337. When do the domestic drivers' hours rules apply?

A. They apply to drivers of goods vehicles on journeys within the UK which are exempt from the EC regulations.

Q338. Are there any exemptions to the domestic drivers' hours rules?

A. Yes. The rules do not apply to drivers of vehicles used by the armed forces, the police and fire brigades, or to drivers who always drive off the public road system. They also do not apply to private driving not in connection with any employment.

In addition, drivers of small goods vehicles not exceeding 3500kg permissible maximum weight and dual-purpose vehicles are only subject to the 10 hour daily driving limit (see Question 355) when used:

- by doctors, dentists, nurses, midwives and vets
- for any services of inspection, cleaning, maintenance, repair, installation or fitting
- by a commercial traveller
- by the AA, RAC or RASC
- for cinematographic or radio and television broadcasting.

Q339. Is there any requirement for drivers to keep records under the domestic drivers' hours rules?

A. Yes. Drivers must keep written records of the hours worked on a weekly record sheet. Operators are expected to check and sign each weekly record sheet.

Tachographs must be fitted and used on all vehicles with a permissible maximum weight of over 3500kg which carry parcels on postal services. The drivers of such vehicles are exempt from the EC drivers' hours rules but must comply with the domestic drivers' hours rules.

Q340. Are there any exemptions from keeping records under the domestic drivers' hours rules?

A. Yes, for drivers of goods vehicles which do not require an Operator's Licence. This exemption does not apply to drivers of Crown vehicles, which would have needed an Operator's Licence if the vehicle had not been Crown property. Drivers of goods vehicles are exempt on any day when they drive for four hours or less and keep within 50km of base. In addition, drivers using an EC calibrated and sealed tachograph are exempt.

Q341. What happens if the domestic drivers' hours rules are contravened?

A. The driver and employer may be liable to a penalty of up to £2500. The Operator's Licence may also be jeopardised.

Q342. What is the position when drivers drive partly under the EC drivers' hours rules and partly under the domestic drivers' hours rules?

A. Where drivers use vehicles which are subject to the EC rules during a day or week in which they also drive vehicles subject to the domestic rules, they may either observe the EC rules all of the time, or observe a combination of both, provided the EC limits are not exceeded when the driver is driving on EC work.

The following points must be considered:

- time spent driving under the EC rules cannot count as an off-duty period under the domestic rules
- driving and other duties under the domestic rules, including non-driving work in another employment, count as attendance at work but not as a break or rest period under the EC rules
- driving under EC rules counts towards the driving and duty limits under the domestic rules
- any EC driving in a week means that drivers must take daily and weekly rest as specified.

Q343. What vehicles are subject to the tachograph regulations?

A. Vehicles used for the carriage of goods, the permissible maximum weight of which, including any trailer or semi-trailer, exceeds 3500kg.

Q344. Are there any exemptions from the tachograph legislation?

A. Yes. The following are exempt in relation to operations anywhere in the EU.

EXEMPTIONS FROM THE TACHOGRAPH REGULATIONS

- Vehicles that are not used on the public road.
- Vehicles with a maximum authorised speed not exceeding 30 km/hour (19.6 mph).

- Vehicles used by: the police, gendarmerie, armed forces, fire brigades, civil defence, drainage or flood protection authorities; the water, gas or electricity services; the highway authorities and refuse collection services; the telegraph or telephone services; the postal authorities for the carriage of mail; the radio or television services, or vehicles used for the detection of radio or television transmitters or receivers; or vehicles which are used by other public authorities for public services, and which are not in competition with professional road hauliers.

- Vehicles used in emergencies or rescue operations.

- Specialised vehicles used for medical purposes.

- Vehicles used for transporting circus and funfair equipment.

- Specialised breakdown vehicles; defined as a vehicle which is specially equipped with some form of crane for lifting/towing purposes for the removal of broken-down vehicles or mobile plant from the point where the breakdown occurred (the exemption does not cover vehicles used to collect cars from auction even if they are unroadworthy).

- Vehicles undergoing road tests for technical development, repair or maintenance purposes, and new or rebuilt vehicles which have not yet been put into service.

- Vehicles used for the non-commercial carriage of goods for personal use.

- Vehicles used for the collection of milk from farms and the return to farms of milk containers or milk products used for animal feed.

In addition, the following are exempt when operating within the UK.

- Vehicles used by agricultural, horticultural, forestry or fishery undertakings for the carriage of goods within a 50km (31 mile) radius of the place where the vehicle is normally based, including local administrative areas, the centres of which are situated within that radius. In the case of fishery undertakings, the exemption applies only to the movement of fish from landing to first processing on land and of live fish between fish farms.

- Vehicles used for the carriage of animal waste or carcasses which are not intended for human consumption. The term "carcasses" includes dead poultry and part carcasses, but not such things as frozen chicken portions, sausages or packaged lamb chops, etc. "Animal waste" is slaughterhouse scraps and offal, but it does not include such things as organic chemical compounds, skins and hides if these are *en route* to be processed into shoe leather.

- Vehicles used for the carriage of live animals from farms to local markets and vice versa, or from markets to local slaughterhouses. The term "animals" includes poultry.

- Vehicles used as shops at local markets or for door-to-door selling, and specially fitted for such uses. Specialised vehicles in this context are vehicles specially constructed or adapted to carry/distribute the commodity being sold, as distinct from a vehicle that could be used for general purposes.

- Vehicles used for: mobile banking; exchange or savings transactions; worship; the lending of books, records or cassettes; cultural events or exhibitions and which are specially fitted for such uses.

- Vehicles with a maximum permissible weight of not more than 7500kg used for the carriage of material or equipment for the driver's use in the course of his or her work within a 50km (31 mile) radius of the place where the vehicle is normally based, provided that the driving of the vehicle does not constitute the driver's main activity.

- Vehicles operating exclusively on islands not exceeding 2300 square km in area, which are not linked to the rest of Great Britain by a bridge, ford or tunnel open for use by motor vehicles.

- Vehicles with a gross vehicle weight, including batteries, of not more than 7500kg, used for the carriage of goods and propelled by means of gas or electricity.

- Vehicles used for driving instruction with a view to obtaining a driving licence, but excluding instruction on a journey connected with the carriage of a commercial load.

- Vehicles operated by the Royal National Lifeboat Institution.

- Vehicles manufactured before 1 January 1947.

- Vehicles propelled by steam.

- Vehicles used by health authorities as ambulances to carry staff, patients, medical supplies or equipment.

- Vehicles used by local authority social services departments to provide services for the elderly or physically or mentally handicapped.

- Vehicles used by HM coastguard and lighthouse services.

- Vehicles used by harbour or airport authorities if the vehicles concerned remain wholly within the confines of ports and airports.

- Vehicles used by British Rail and other transport authorities when engaged in maintaining railways.
- Vehicles used by the British Waterways Board when engaged in maintaining navigable waterways.
- Tractors used exclusively for agricultural and forestry work.

Q345. **What are the employer's responsibilities regarding drivers' hours and tachograph regulations?**

A.   Employers are required to:

- issue a sufficient number of tachograph record sheets to drivers bearing in mind that the charts are personal to the drivers, and consider the length and period of the journey, and the possible need to replace charts which may be damaged or which may be taken by an authorised enforcement officer
- ensure that drivers return the charts within 21 days
- keep the tachograph charts for a period of at least one year, and produce them when requested to do so by an authorised inspecting officer.

Both employers and drivers are responsible for seeing that the tachograph equipment functions correctly and that the seals remain intact.

In the event that police officers wish to see the drivers' records at the operator's premises they cannot remove them, but may inspect and copy them or may require that the records be taken to the office of the licensing authority.

Q346. **What happens in circumstances where a required vehicle does not have a tachograph fitted or where a tachograph is not being used in accordance with the regulations?**

A.   The employer can face a fine of up to £5000 on conviction.

Q347. **Is there any duty on employers to check the tachograph charts?**

A.   Yes. The regulations require employers to carry out "periodic" checks of tachograph records to ensure that the drivers' hours rules are being complied with. Where breaches of the rules are discovered, employers have a duty to take steps to ensure that they do not recur.

Q348. **What are the duties placed on drivers?**

- Drivers are required to keep the tachograph running continuously from the time that they take over the vehicle until they are relieved of their responsibilities for it.

- Drivers are required to use the mode switch on the tachograph enabling driving time, breaks from work and rest periods, and other periods of work and attendance at work, to be recorded separately and distinctly.

- Dirty or damaged charts must not be used. Where a chart containing recordings has been damaged, drivers must attach the damaged chart to the chart used to replace it.

- Drivers are required to keep with them used charts for the current week and for the last day of the previous week in which they drove. A driver who had been on holiday or absent from work for a period of time would, therefore, carry the chart for the last day on which he or she drove prior to the holiday or period of absence.

- Drivers are required to hand completed tachograph charts in to their employer within 21 days.

- If an enforcement officer retains a tachograph chart, drivers should obtain a receipt, which can be shown to another enforcement officer if required and which can be handed to employers in place of the chart.

Q349. **When drivers recommence driving after a holiday, must they carry with them the chart covering the last day on which they drove before the holiday?**

A.   Yes. The reference to "previous week" in Regulation 3821/85 is to the last week in which driving was undertaken. In practice, this means, for example, that if a driver was on holiday for a fortnight, he or she must have with him or her during the first week back, the chart for the last driving day in the last week that he or she drove, even though in this example, there would be a gap of over a fortnight.

Q350. **What happens when drivers are away from their vehicle and unable to operate the tachograph?**

A.   On leaving the vehicle, drivers may make a manual entry on the reverse of the chart or may elect to leave the chart in the tachograph and select the appropriate driver mode. It is important to remember, however, that it is an offence to leave a chart in the tachograph for longer than 24 hours.

Q351. Does other work after a driver has finished driving have to be recorded on the tachograph chart?

A.   Yes. The High Court has ruled that overtime worked in the yard after a driver has completed driving for that day has to be recorded. Once the decision is made to work overtime, the driver is not freely able to dispose of his or her time for it to be counted as a rest period under EC Regulation 3280/85.

Q352. What information do drivers have to enter on the centre field of a tachograph chart?

A.   Drivers are required to enter the following information.

- At the beginning of the use of the tachograph chart, the driver's surname and full first name. No initials, abbreviations, shortened names or nicknames are acceptable. If more than one driver with the same name uses a vehicle, some way of distinguishing them must be found, such as the use of a middle name.
- The date and place where the use of the chart begins and the place where such use ends. Place names should be written in full, without abbreviations. The use of words such as "depot", "base" and "home" are not acceptable. Exact locations should be used, rather than general names, such as "London".
- The registration number of each vehicle to which the driver is assigned, both at the start of the first journey recorded on the chart and then, in the event of a change of vehicle, during the use of the chart. Fleet numbers should not be used in place of registration numbers. Where there is a change of vehicle, the time of the change from one vehicle to the next must be written on the chart.
- The odometer reading at the start of the first journey recorded on the chart, at the end of the last journey recorded on the chart and, in the event of a change of vehicle during the working day, the odometer readings at the beginning and end of the use of each vehicle. All the numbers on the odometer reading should be entered on the chart.

Q353. How often should tachographs be inspected and calibrated?

A.   Tachographs must be inspected at a tachograph calibration centre approved by the Department of Transport every two years to ensure that the equipment is working properly. The two year inspection is due either two years after the date shown on the installation plaque or two years after the date shown on the two yearly inspection plaque.

Tachographs must be recalibrated by an approved calibration centre every six years. Recalibration is due six years after the date shown on the installation plaque. The installation plaque should be located beside the equipment or on the equipment itself, so as to be clearly visible. When the tachograph is recalibrated, and the installation plaque replaced, a two year inspection must follow.

Where a repair to a vehicle involves the recalibration and resealing of the tachograph, the inspection and recalibration periods apply from that date.

Q354. What happens when a tachograph is faulty or the seals are broken?

A.   In the event of a breakdown or faulty operation of the tachograph, employers must have it repaired at an approved calibration centre as soon as the vehicle has returned to base. If the vehicle is unable to return to base within seven days, including the day of the breakdown or discovery of the defect, the repair must be carried out before it returns to base.

Whilst the tachograph is unserviceable, or operating defectively, drivers are required to make a temporary record either by writing on the chart or using a temporary sheet. The temporary sheet must be attached to the chart and must contain all the information that has not been correctly recorded by the tachograph.

Q355. What are the daily driving limits under EC and domestic rules?

A.   Under the EC rules, the daily driving limit is nine hours; this can be increased to 10 hours not more than twice a week. A "day" is regarded as the period between two daily rest periods or the period between a daily rest period and a weekly rest period.

Under the domestic rules, drivers must not drive for more than 10 hours in a day. In addition, drivers must not be on duty for more than 11 hours in any working day. Drivers are exempt from the daily duty limit on any working day when they do not drive or if they do not drive for more than four hours on each day of the week.

Q356. How long can drivers drive under the EC rules before they have to take a break?

A.   Drivers have to take a break of 45 minutes after 4½ hours continuous or accumulated driving, unless they begin a daily or weekly rest period. Alternatively, drivers may take a combination of breaks of at least 15 minutes distributed over the 4½ hour driving period or at the end of that period.

During such breaks drivers are not allowed to carry out other work.

Waiting time and time not devoted to driving, eg spent on a vehicle in motion, on a ferry or train, is not regarded as other work for this purpose.

Where a driver has taken 45 minutes' break, either as a single break or as several breaks of at least 15 minutes during or at the end of the 4½ hour period, the calculation should begin afresh, without taking account of the driving time and breaks previously completed by the driver.

Q357. Is there a limit on the number of hours drivers can drive in a fortnight under the EC rules?

A.   Yes. Drivers are not permitted to drive for more than 90 hours in a fortnight.

Q358. What is the definition of a working "day" and a working "week"?

A.   Under the EC Rules, *days* are defined as successive periods of 24 hours beginning with the resumption of driving after the last weekly rest period. "Each period of 24 hours" in Article 8(1) of EC Regulation 3820/85 means any period of 24 hours commencing at the time when the driver activates the tachograph following a weekly or daily rest period. The calculation of the driving period begins at the moment when the driver sets in motion the tachograph and begins driving.

Under the domestic rules, a day is defined as any period of 24 hours.

Under both sets of rules, a *week* is defined as the period between midnight on Sunday and midnight on the following Sunday.

Q359. What are the daily rest requirements under the EC rules?

A.   Drivers must take a daily rest period of not less than 11 consecutive hours, which may be reduced to nine hours on not more than three occasions in any one week. Any reduction must be compensated for by an equivalent period of rest being added to a daily or weekly rest period before the end of the following week. Alternatively, a daily rest period of 12 hours may be taken split into two or three periods, one of which must be of at least eight hours and none of which must be less than one hour. Where the daily rest is taken in two or three separate periods, the calculation must commence at the end of the period of not less than eight hours.

Where vehicles are double manned, each driver is required to have a daily rest period of not less than eight hours during each period of 30 hours.

Daily rest periods may be taken in the vehicle, provided that it is equipped with a bunk and it is stationary. This prevents drivers of double-manned vehicles taking their rest periods whilst their partners are driving.

Q360. **What are the weekly rest requirements under the EC rules?**

A.   Under the EC rules, a driver must take a weekly rest period after six daily driving periods. However, the weekly rest period may be postponed until the end of the sixth day if the total driving time over the six days does not exceed the maximum corresponding to six daily driving periods, normally 56 hours.

Drivers must take a weekly rest period of 45 consecutive hours. That period can be reduced to 36 consecutive hours if the rest is taken where the vehicle or driver are normally based, and to 24 consecutive hours if taken elsewhere. Reductions taken must be compensated for *en bloc* in conjunction with a daily or weekly rest period before the end of the third week after that in which the reduction occurs.

A weekly rest period which begins in one week and continues into the following week can be attached to either of those weeks.

Weekly rest periods may be taken in the vehicle, provided that it is equipped with a bunk and it is stationary.

Q361. **What happens when drivers are on a ferry boat or train?**

A.   Where part of a vehicle's journey is by ferry boat or train, the following rules apply.

- The rest period may be split, with part being taken on land preceded or followed by a period of rest taken on the boat or train. The rest period must only be interrupted once, and two hours must be added to the total rest time. The interruption to the rest period must be as short as possible, and must be no more than one hour before embarkation or one hour after disembarkation, customs formalities being included in the embarkation and disembarkation procedures.

- During both parts of the rest period, the driver must have access to a bunk or couchette.

- Any time spent on board not counted as part of the rest period is regarded as a break from driving or other work.

- The rest period or break on board ship or train is deemed to have started once drivers are free to leave their vehicles, and it continues until they are instructed to rejoin their vehicles prior to disembarkation.

Q362. **What happens in emergencies?**

A.   Provided that road safety is not jeopardised, and in order to reach a suitable stopping place, drivers may depart from any of the EC drivers' hours rules only to the extent that is necessary to ensure the safety of persons, the vehicle or its load. Drivers are required to indicate the nature of any such departure, and the reason for it, on their tachograph charts.

Under the domestic drivers' hours rules, the limits may be exceeded when there are:

- events which cause or are likely to cause danger to life or health of one or more individuals or animals; or a serious interruption in the maintenance of public services for the supply of water, gas, electricity or drainage or of telecommunications or postal services; or serious interruption in the use of roads, railways, ports or airports
- events which are likely to cause such serious damage to property, that it becomes necessary for immediate action to be taken to prevent such danger or interruption occurring or continuing, or damage being caused.

Q363. Do the EC regulations prohibit bonus payments to drivers?

A. The regulations prohibit payments to drivers in the form of bonuses or wage supplements which are related to the distance travelled and/or the amount of goods carried, unless the payments are of such a kind that do not endanger road safety.

SPEED LIMITERS

Q364. **Where can the law in relation to speed limiters be found?**

A.   The UK law is contained in Regulations 36A, 36B, 70A and 70B of the **Road Vehicles (Construction and Use) Regulations 1986** (SI 1986 No.1078). The regulations will eventually have to be changed so that the minimum requirements of EC Directive 6/1992 are implemented.

Q365. **What are the main provisions of EC Directive 6/1992?**

A.   The Directive applies to motor vehicles with at least four wheels and a maximum design speed exceeding 25 km/h (15.5 mph), with a maximum weight exceeding 12 tonnes, first registered after 1 January 1994. Such vehicles can only be used on a road if the speed limiter is set in such a way that the vehicle's speed cannot exceed 90 km/h (55.8 mph) and the maximum speed must be set at 86 km/h (53.3 mph).

Vehicles first registered between 1 January 1988 and 1 January 1994 became subject to the EC rules from 1 January 1995, unless they are used exclusively on national transport operations, when the implementation date will be 1 January 1996.

Q366. **Are there any exceptions to the EC rules?**

A.   Yes. The following vehicles are exempt:
- those used by the armed forces, civil defence, fire and other emergency services or forces responsible for maintaining public order
- those used for scientific tests on roads
- those used only for public services in urban areas, ie buses.

Q367. **Which vehicles (under UK law) require speed limiters to be fitted?**
- Any vehicle with a maximum gross weight over 7500kg, first used on or after 1 August 1992, and which is capable of exceeding 97 km/h (60 mph) on the level without a speed limiter, must *not* be used on the road unless it is fitted with a speed limiter.
- From 1 August 1993, any vehicle which has a maximum gross weight of over 16,000kg, first used on or after 1 January 1988, and which is capable of exceeding 97 km/h (60 mph) on the level without a speed limiter and is either:

- a rigid vehicle constructed to draw a trailer and the difference between its plated gross weight and gross train weight exceeds 5000kg, or

- constructed to form part of an articulated vehicle

must not be used on the road unless it is fitted with a speed limiter.

Q368. Which vehicles are exempt?

A. The following vehicles are exempt:

- vehicles being taken to a place where a limiter is to be installed or calibrated

- vehicles owned by the Secretary of State for Defence and being used for military purposes

- vehicles being used for military purposes whilst driven by a person subject to the orders of the armed forces

- vehicles being used for fire brigade, police or ambulance purposes

- vehicles completing a journey during the course of which the limiter has accidentally ceased to function.

Q369. Do speed limiters have to meet a particular specification?

A. Yes. Speed limiters must comply with British Standard BS AU 217 or with a standard or technical regulation of, or recognised by, another EU Member State and which offers equivalent guarantees of safety, suitability and fitness for purpose.

In addition, speed limiters must be:

- calibrated to a set speed not exceeding 97 km/h (60 mph) (vehicles up to 12,000 kg) or 90 km/h (56 mph) for vehicles over 12,000 kg

- sealed by an authorised sealer (see Question 372) to protect them against improper interference, improper adjustment or interruption of their power supply

- maintained in a good and efficient working order.

Q370. What are the requirements of British Standard BS AU 217?

A. Part 1a of the Standard requires the speed-sensing device of the speed limiter to be accurate to within ± 2.4 km/h (1.5 mph) of the vehicle's actual speed throughout the limiter's effective range. The sensed speed of the vehicle must not differ from the set speed marked on the vehicle by more than 3.2 km/h (2 mph).

When the vehicle is running at its set speed with the accelerator at maximum travel and is subject to an accelerating force due to a down gradient, the limiter must control speed by restricting engine power to the speed limiter's minimum engine power setting. A malfunction of the speed limiter must not result in an increase of engine power above that caused by the accelerator position.

The supplier of the limiter must provide documentation describing checking and calibration procedures and it must be possible to check the set speed while the vehicle is stationary.

Q371. How can you tell whether or not a vehicle has been fitted with a speed limiter?

A. Vehicles that have speed limiters fitted must also have a plate fixed in a prominent and readily accessible position in the driving compartment. This plate is supplied by the authorised sealer. The plate must be clearly marked with the speed limiter setting.

The requirements in relation to sealing and the fixing of a plate do not apply to speed limiters fitted to vehicles before 1 August 1992 and used before that date or to speed limiters fitted outside the UK.

Q372. Who is regarded as an "authorised sealer"?

A. An authorised sealer is a person, firm or company authorised by the Secretary of State for Transport to seal speed limiters. An authorised sealer can charge for sealing limiters and the charges are not controlled by legislation.

Q373. Is it an offence to drive a vehicle on the road when the speed limiter is not functioning?

A. Yes. The driver and the employer may be liable to a penalty of up to £2500. The only exceptions are when a journey is being completed during which the speed limiter has accidentally ceased to function or when the vehicle is being taken to a place where the speed limiter is to be repaired or replaced.

INDEX

The index covers the whole book except the numerical list of questions on pages 1–16. Index entries are to page numbers. Alphabetical arrangement is word-by-word, where a group of letters followed by a space is filed before the same group of letters followed by a letter, eg "tax discs" comes before "taxation". In determining alphabetical arrangement, initial articles and prepositions are ignored. Publications mentioned in the text are printed in italics.